Ecological Research Monographs

More information about this series at http://www.springer.com/series/8852

Ryo Sakurai

Human Dimensions of Wildlife Management in Japan

From Asia to the World

Springer

Ryo Sakurai
College of Policy Science
Ritsumeikan University
Osaka, Japan

ISSN 2191-0707 ISSN 2191-0715 (electronic)
Ecological Research Monographs
ISBN 978-981-13-6331-3 ISBN 978-981-13-6332-0 (eBook)
https://doi.org/10.1007/978-981-13-6332-0

Cover illustration: Landscape of a village in Kami town of Hyogo Prefecture, one of the research sites studied in this book. Photo by Ryo Sakurai.

This Springer imprint is published by the registered company Springer Nature Singapore Pte Ltd.
The registered company address is: 152 Beach Road, #21-01/04 Gateway East, Singapore 189721, Singapore

Preface

On August 26, 2016, I was at St. Petersburg airport in Russia, waiting for the airplane with Dr. Ken'ichi Nakagami (former professor of Ritsumeikan University and the University of Tokyo). We had just taken part in the 11th International Conference on the Environmental Management of Enclosed Coastal Seas (EMECS11) in St. Petersburg and were on our way back to Japan. In the airport, I was catching up with some academic work, which is something I do whenever I have some free time. Dr. Nakagami saw me writing a paper. He smiled, and asked, "Are you still writing papers? Why are you writing papers?" At first, I did not understand what he meant, as I believed that writing papers and publishing them in academic journals was the most important and meaningful thing that a researcher could do. Dr. Nakagami continued, "You should write a book, not a paper." He went on to explain that writing a book has different meaning and value to writing a paper, as it contributes to society as a whole, not only to academia.

That was the moment that I understood the importance of writing books. Until then, I thought young researchers should focus on writing as many papers as possible. However, there were few books published in the field of human dimensions of wildlife management in Asia. Human-wildlife conflicts are a worldwide issue, and the concepts of the human dimensions approach were not well understood or utilized in my home continent. I knew that somebody had to write a book, and perhaps I had been avoiding doing it myself. And Dr. Nakagami's encouragement was what it took – I cannot thank him enough. As soon as I came back to Japan, I wrote a book proposal and sent it to Springer.

This is one of the first books published in English that explains about human dimensions of wildlife management in Asia, specifically in Japan. I discussed, in this book, how human dimensions can be utilized in Asia, how to conduct effective research in different cultural setting, and how to implement wildlife management successfully. It is my wish that this book contributes to our understanding of how human dimensions approach could help mitigate human-wildlife conflicts and foster effective decision-making.

Osaka, Japan — Ryo Sakurai

Acknowledgment

I first would like to express my deepest gratitude to Dr. Susan K. Jacobson, my advisor and mentor at the University of Florida, for both my Master's and Doctoral programs. She taught me all the important elements of the human dimensions discipline, including how to design and conduct human dimensions research and even how to behave in the field. I would therefore like to dedicate this book to Susan. I also would like to thank my colleagues/labmates in the graduate school, Dr. Stuart Carlton and Dr. Dara Wald, for their stimulating ideas about research and their help in adjusting to a new environment in the USA. Second, special thanks go to Dr. Gouhei Ueda, the co-researcher who taught me so much about the situation relating to wildlife management and human-wildlife conflicts in Japan. Third, I would like to thank the two professors who have helped guide my research and my career: Dr. Hiromi Kobori for her ongoing support, for providing me with an office while I was a graduate student in Japan, for encouraging me to take part in academic conferences, and for teaching me the concept of citizen science; and Dr. Hiroyuki Matsuda, who taught me how social scientists could collaborate with ecologists to achieve better natural resource management and who was generous enough to let me join his Man and Biosphere Reserve projects in Japan. Fourth, and finally, I would like to thank Springer for their support, especially Ms. Kripa Guruprasad and Ms. Fumiko Yamaguchi.

On a personal note, I would like to thank my parents for their tireless support. And, finally, my sincere gratitude to my wife, Masako, for all of her support and encouragement.

Contents

Chapter 1
Introduction: New Perspectives of Wildlife Management

Faced with numerous environmental concerns, such as deforestation, desertification, and overfishing, human beings have begun to realize that there is no such thing as "unlimited resources" and that we must live sustainably if we intend to continue calling planet Earth our home. With this in mind, conservation of the natural environment and its biodiversity has become a globally shared goal over the last few decades, as shown in the Aichi Biodiversity Targets and Sustainable Development Goals, etc. Wildlife have been hunted historically for food, fur, and other uses and are sometimes overhunted to extinction (such as the passenger pigeon [*Ectopistes migratorius*] and the dodo [*Raphus cucullatus*]). Large carnivores, like Japanese wolves (*Canis lupus hodophilax*), have been seen by man as a threat to human lives or livestock and have been hunted to extinction for this reason in many countries around the world. On the other hand, researchers in ecologically based academic disciplines have compiled data and provided evidence that each species on Earth plays a specific role in its ecosystem and constitutes a vital part of the biodiversity within. Many countries have set forth laws restricting the overhunting of species, creating policies not only to conserve threatened species but threatened ecosystems as well. Establishing national parks and/or preservation areas has been thought of as an effective approach to conserving natural environments, and many countries have taken this approach. As of 2016, Japan had established 34 national parks covering about 1% of the total land in the country.

However, some people have realized that establishing conservation areas is not always effective in terms of sustainably managing natural resources. For example, in some areas, residents have claimed their rights to continue harvesting resources with which they have been historically associated, resulting in illegal entry to the area and poaching of endangered or threatened species. Hunting restriction laws could work to prevent illegal and unsustainable hunting to some degree; however, in some countries, these laws have led to dangerous rapid population increases in certain species, such as deer and wild boar, causing serious agricultural damage. As a result, these populations must be controlled, not protected. Meanwhile, in countries like Japan and the USA, the numbers of hunters have been decreasing in recent

R. Sakurai, *Human Dimensions of Wildlife Management in Japan*, Ecological Research Monographs, https://doi.org/10.1007/978-981-13-6332-0_1

years, leaving abundant concerns over who will control the overpopulated species and how will it be accomplished. In situations where large carnivorous species, such as tigers or bears, have attacked livestock or people, wildlife managers are facing challenges in reaching consensuses for actions among stakeholders. Local residents often want to remove or kill the animals in question living around human settlements, while animal rights advocates claim that all wildlife species, including the carnivores responsible for the attacks, have the right to live in their natural habitats. At the same time, tourist industries are most concerned with utilizing those species to attract visitors.

Faced with these issues and challenges, people have started to recognize that wildlife management is not only about "managing wildlife" but is also about "managing people." To implement an effective wildlife management strategy, managers must consider how to balance the needs, demands, and requests of the various stakeholders involved. This became known as "New Perspectives of Wildlife Management," and in an effort to reflect the various needs and backgrounds, diverse research was conducted regarding public opinions and attitudes toward wildlife. In the USA in the 1970s, these efforts led to the establishment of a new academic discipline, "human dimensions of wildlife management (HDW)." Since that time, research has been compiled under the discipline, and theorizing models and schemes applicable in the field have been tested. Historically, however, a majority of the HDW research has been conducted in North America, whereas very few studies have been implemented on other continents, such as Asia. The question of how the HDW tools, ideas, and methods that have developed in the USA could be applied to other regions of the world having completely different cultural, social, and historical backgrounds remains mostly unexplored.

I will never forget the comments made to me by a wildlife manager in a rural village in Japan when I first began my research based on HDW theories. I proposed to conduct a survey of all villagers, to understand their perceptions about local wildlife issues. He promptly replied by saying, "…in these regions, all the issues in the villages are decided by village chiefs. Villagers just follow what village chiefs decide; therefore, perception of each resident does not matter because it does not affect the decision. With this in mind, what's the purpose of conducting a survey of all the villagers?"

In another case, in a different prefecture, I also remember very clearly what one researcher said to me when I proposed forming a focus group interview to understand stakeholders' opinions about wildlife issues. The researcher said, "In Japan, especially in such rural areas, people usually do not speak up in a public meeting. People basically stay calm and quiet; therefore, a focus group interview will not work, and even more so, the tools and methods developed in the U.S. will not work in Japan."

The main topic of this book is whether HDW ideas, tools, and approaches will work in a different culture and, if so, in what sense? The comments by the manager and the researcher above imply that a certain number of people believe that such tools and methods will not work effectively in a different cultural setting. It is true that the lifestyles and beliefs in Asia and North America are different. Asian people (from Japan, China, Korea, etc.) are said to be collectivistic, caring about what others think and say, whereas Americans, for example, are believed to be more individualistic. As we start to understand that the cultural and social characteristics within each country affect how wildlife is managed there, we need to prepare new methods for wildlife management; management procedures should be based on the specific cultural, social, and historical backgrounds of the region where the management is implemented. This brings us to another "New Perspectives of Wildlife Management." While we have been experiencing an unprecedented globalization in which goods, information, and people move rapidly across countries and continents, there are still cultural and social differences that define the identities of each country and even each region within countries. As long as each country has its own culture, the usability and reliability of HDW as an academic discipline will be challenged by the same question: "does that work in a different cultural setting?"

In this book, I discuss how HDW can be utilized in Asia, the necessary considerations to conduct effective research in different cultural settings, and overall how to implement wildlife management successfully.

In the second chapter, I explain the basic concepts, including definitions, objectives, and the history of the field of human dimensions of wildlife management. Since this field was founded and developed in North America, the first part of the chapter primarily focuses on case studies in North America; however, as human-wildlife conflicts occur globally, this chapter also discusses the situation outside of North America, including Europe, Africa, and Asia.

In the third chapter, the history of wildlife management in Japan is explained. This chapter covers the evolution of human-wildlife relationships in Japan and how law and policies regarding wildlife management have changed over time. In addition, I discuss the current status of human-wildlife conflicts in Japan, such as increases in certain species (e.g., deer and wild boar), and common implications, like agricultural damage.

In the fourth chapter, I discuss one of the most serious current human-wildlife conflicts in Japan, black bear issues, and I describe the human dimensions studies that have been conducted to solve these problems. This chapter begins with an explanation of human-bear conflicts in Japan, where I specifically focus on media portrayal of these issues. I move, then, to look at case studies of the human dimensions research conducted in one specific area of Japan, the Tajima region (the northern part) of Hyogo Prefecture, where I discuss the results of several studies conducted to understand local residents' and stakeholders' perceptions of black bear issues. The final portion of this chapter explains the community bear education seminar implemented for local residents and how this seminar contributed to changing participants' attitudes regarding bear issues, as well as fostering behaviors aimed at preventing bear damage.

The fifth chapter explains human dimensions studies conducted in Central Japan, Tochigi Prefecture. In this particular prefecture, participatory model district programs were implemented with support from the prefectural government to reduce damage by wildlife, such as macaques, wild boar, and deer. The first part of this chapter explains results from the human dimensions research conducted to understand the significance and impacts of these model district programs in terms of changing residents' attitudes and behaviors, as well as reducing human-wildlife conflicts in the region. The second part of the chapter focuses on a study that tested specific social-psychological theories in understanding and encouraging residents' behavioral intentions regarding wildlife issues.

The sixth chapter explains how human dimensions ideas and approaches could be introduced and utilized throughout Asia. Tips and suggestions are introduced for conducting human dimensions studies in Asian countries based on the case studies explained in Chaps. 4 and 5. Some useful tools also are explained, such as the Potential for Conflict Index and the wildlife acceptance capacity, which are effective tools for demonstrating these types of results to general public.

The seventh chapter explains how the human dimensions field could be used to educate the Asian people in wildlife management concepts. Based on the author's teaching experiences of the human dimensions of wildlife management in Japanese universities, this chapter explains the importance of training Asian students to become professionals in the human dimensions field.

The eighth and final chapter discusses the potential of human dimensions for solving human-wildlife conflicts, not only in Asia but also on a global scale. Many countries are beginning to face similar problems as those confronted by Japan and other Asian countries, such as a shrinking and aging society (e.g., England) and a decrease in the hunting population (e.g., USA). On the other hand, in some parts of Asia, the rapidly growing human population is threatening wildlife. The situation regarding the burgeoning human population does exist in other parts of the world as well, such as in some African countries; therefore, Japan and other Asian countries should become model cases for introducing human dimensions approaches into management strategies for solving human-wildlife conflicts. This final chapter also introduces ideas and methods for citizen science programs and explains some of the case studies where local residents have joined or initiated scientific research to better understand the wildlife populations and their activities around villages. We believe that citizen science projects could be an effective method for implementing collaborative wildlife management among local residents, researchers, and governments. In addition, this chapter discusses the potential for broadening the theme of human dimensions, not only to the management of wildlife but also to the management of natural resources and biodiversity. For example, this chapter introduces case studies in which HDW ideas and tools have been applied to coastal zone management.

Chapter 2
What Is "Human Dimensions"?

Abstract Human dimensions of wildlife management (HDW) is an academic discipline founded and developed in the USA. HDW could be explained simply as a practical science for implementing wildlife management utilizing a social science approach. HDW aims to realize effective wildlife management practices by reflecting the needs and opinions of citizens and stakeholders into the adopted policies. The significance of HDW is that it enables managers to increase fairness, balance, and legitimacy in the decision-making process by including stakeholder' needs and opinions. HDW has begun to be recognized outside of North America, particularly in Europe and Africa; however, it remains a relatively new field, struggling to be accepted as an academic discipline. Meanwhile, HDW is currently not well known in many Asian countries. I establish here the importance of adopting these methods in Asian countries and the reasons behind their hesitancy for doing so.

2.1 Human Dimensions of Wildlife Management in North America

Human dimensions of wildlife management is an academic discipline founded and developed in the USA. Decker et al. (2012) pointed out that wildlife management, conventionally, is said to be composed of three aspects: humans, wildlife, and habitats. Therefore, "everything in the wildlife management system that is not directly about wildlife and habitats is about humans" (Decker et al. 2012, p. 3), and these things are considered the range of human dimensions. One of the first textbooks in this field, *Human Dimensions of Wildlife Management in North America*, defined HDW as a discipline to study "how people value wildlife, how they want wildlife to be managed, and how they affect or are affected by wildlife and wildlife management decisions" (Decker et al. 2001, p. 3). It is also explained that "human dimensions covers a broad set of ideas and practices, including economic and social values, individual and social behavior, public involvement in management decision making, and communication" (Decker et al. 2001, p. 3). Based on several definitions of HDW, this field could be explained as a practical science for implementing

R. Sakurai, *Human Dimensions of Wildlife Management in Japan*, Ecological Research Monographs, https://doi.org/10.1007/978-981-13-6332-0_2

wildlife management utilizing a social science approach. HDW aims to realize effective wildlife management by reflecting the opinions and needs of citizens and stakeholders into adopted policies (Manfredo and Vaske 1996).

Although wildlife management has been explained as 90% people management and 10% wildlife management (Fazio and Gilbert 1986, p. 3), research studies regarding wildlife management have been performed, traditionally, by ecologists and biologists who primarily studied only the wildlife itself and/or their habitats. Previously, wildlife managers had to make decisions based on the ecological and biological aspects of wildlife as early as the 1930s, where state agencies collected mostly biological information from hunters (Decker et al. 2001). One of the first human dimensions studies was a national survey conducted in 1955 by the US Bureau of Sport Fisheries and Wildlife (known today as the US Fish and Wildlife Services) that was intended to track "wildlife-associated recreation participation and economic expenditures" for Americans (Manfredo 2009).

In the USA, the Public Trust Doctrine declared that wildlife property is owned by no one, but treated as public property, and held in trust by the government for the benefit of citizens (Organ and Batcheller 2009). This law establishes strong reasons for why governments and wildlife managers must understand the needs and demands of the public to make appropriate decisions regarding wildlife management. This becomes especially true as certain species populations begin to increase, resulting in both increased and diversified human-wildlife interactions and, therefore, increasing the necessity for studying human dimensions to help implement effective management practices (Decker et al. 2001).

The significance of HDW is that it enables managers to increase the fairness, balance, and legitimacy in decision-making processes by reflecting stakeholder' needs and opinions. Meanwhile, HDW also contributes to efforts for changing policies from only coping with difficult wildlife problems to actually implementing measures of precaution (Decker et al. 2001). Two cases that have contributed to the development of HDW were the management of white-tailed deer (*Odocoileus virginianus*) in New York and the reintroduction of wolves (*Canis lupus*) into Michigan and Yellowstone National Park.

In New York, debate regarding the management of white-tailed deer ended with hunters bringing a lawsuit against managers. In the USA, traditionally, wildlife management was implemented by experts with authority, in a top-down approach, without involving stakeholders in the management process. In the 1960s, state biologists set a goal for controlling the population of white-tailed deer in the Adirondack Region of New York; however, hunters could not agree to the goal. Hunters worked through local representatives and the state government to remove wildlife managers in the region. This led the government to realize the importance of gaining stakeholders' perceptions and opinions before making policy decisions by conducting systematic and scientific surveys. The state government gradually shifted wildlife management processes from a top-down approach to a community-based approach, and in the 1990s, the New York State Department of Environmental Conservation formed a citizen committee that facilitated a consensus agreement regarding objectives for the deer population (Decker and Chase 1997; Leong and Haigh 2007; Decker et al. 2001).

The second case mentioned above involves the reintroduction of wolves. In the 1970s, the Natural Resources Department for the state of Michigan reintroduced wolves without entertaining the public opinions toward wolves and without reaching a social consensus on the reintroduction policy. As a result, all four wolves that were released were killed within 8 months. Based on this experience, when a reintroduction of wolves was proposed for Yellowstone National Park, the government conducted a survey of local residents and tourists to better understand their concerns toward this policy. Results of the survey were utilized both by the government and NGOs to conduct outreach programs for stakeholders who had negative opinions about the reintroduction, as well as to host meetings where local residents and managers/experts could communicate and discuss the policy with concerned citizens. The survey and the public outreach are heavily credited with the successful reintroduction of wolves into the park in 1995 (Decker et al. 2001; Jacobson 2009), resulting in both a more stable ecological system and a strong public interest in the park.

In the USA, wildlife management agencies and universities started to establish programs specifically for teaching about the human dimensions of wildlife management in the 1970s. By the end of the 1990s, the specific quarterly journal that compiles studies on this field, *Human Dimensions of Wildlife*, was established. Currently, studies on HDW are quite diverse, including risk perception, communication theories, social trust, and other concepts and theories, and all have been developed and tested by the professionals in this field.

Reviews of papers published in the academic journal, *Human Dimensions of Wildlife*, from 1996 to 2006 revealed that the majority of papers (85%) were quantitative studies using statistical analyses. The rest of the papers (15%) were qualitative in nature, including interviews and focus group discussions. Two-thirds of the quantitative papers used significance testing with a null hypothesis. Meanwhile, more than half of the papers (65%) utilized social-psychological theories, whereas 39% used other social science approaches (such as sociology), and 12% utilized economic concepts (such as using a method of WTP [willingness to pay]). Based on this review, the majority of the HDW studies, at least from 1996 to 2006, were quantitative and statistically vetted research studies using social-psychological theories (Vaske et al. 2006).

Located adjacent to the USA, human dimensions research has been conducted actively in Canada as well. Following the example of the quinquennial (every 5 years) US national hunting and fishing survey, the Canadian Wildlife Service implemented similar national surveys in 1981, 1987, 1991, and 1996, revealing a trend in citizen participation (Decker et al. 2012, p. 208; Gray et al. 1993). One of the classical case studies that actively engaged stakeholders to sustainably co-manage wildlife species took place in Quebec, Canada, in the 1980s regarding the declining beluga whale population (*Delphinapterus leucas*). The Canadian Department of Fisheries and Oceans contacted an organization of Inuit hunters, allowing the hunters to discuss appropriate actions and then implement a sustainable hunting plan (Decker et al. 2012, p. 141).

2.2 Human Dimensions of Wildlife Management Outside of North America

2.2.1 Europe

In Europe, large carnivores, such as brown bears (*Ursus arctos*), Eurasian lynx (*Lynx lynx*), gray wolves (*Canis lupus*), and wolverines (*Gulo gulo*), have shown increases in their populations, expanding widely across the continent (Chapron et al. 2014). This has led to increased human-wildlife interactions and conflicts and has generated the necessity for incorporating human concerns into policies that seek the coexistence of people and wildlife. Since the 1970s, there have been an increasing number of studies conducted in Europe regarding the human dimensions of wildlife resulting in related published documents: 300 published studies after 1994 in Europe (Glikman and Frank 2011). One of the well-studied themes is public attitudes toward bears and wolves. Dressel et al. (2014) implied that people's attitudes toward bears have become more positive, whereas that toward wolves have become relatively negative as an overall trend from 1976 to 2012. However, while the importance of incorporating the human dimensions of wildlife into management issues is finally getting some recognition in Europe, overall the field and its concepts remain relatively obscure there and are struggling to find acceptance as an academic discipline or a decision-making tool (Glikman and Frank 2011). While several HDW-related studies have been conducted in Europe, reports indicate that most of them have been conducted by natural scientists, resulting in studies that lack consistency in conceptualization and measurements (Dressel et al. 2014; Johansson et al. 2016). In addition, many of these studies are one-shot case studies with no follow-up, making it difficult to apply the findings to decision-making processes (Glikman and Frank 2011). These challenges must be clarified and overcome for HDW to be recognized for its significance and effectiveness and ultimately to be accepted as an academic discipline with force in Europe. Meanwhile, it should be noted that European wildlife experts of the Large Herbivore Network (LHNET) and the Large Carnivore Institute (LCIE) have begun recently to communicate the importance of interdisciplinary approaches to wildlife management, including human dimensions measures (Johansson et al. 2016). Other progress, such as HDW being included as a credit-earning course in a Master's program at an Italian university, is important for spreading the word about the critical impacts made by HDW, although the number of universities teaching HDW modules is still limited.

2.2.2 Africa

Africa is well known for being home to incredible numbers of diverse megafauna. Historically, because of the large presence of wildlife, efforts have been made to conserve wildlife and to protect from negative human-wildlife interactions.

Establishing national parks has been one successful method to accomplish these goals in Africa, where species like wild pigs (*Potamochoerus* spp.), elephants (*Loxodonta africana*), hippopotamuses (*Hippopotamus amphibius*), and various primate species commonly cause agricultural damage (Osborn and Hill 2005). Meanwhile, the presence of large carnivores, such as lions (*Panthera leo*), means that human-carnivore conflicts are not unusual throughout Africa (Thirgood et al. 2005). Historically, predators, including the large carnivores, have been perceived as vermin and have been hunted throughout Africa, even in the national parks and reserves, as late as the 1970s (Woodroffe et al. 2005; Marker and Boast 2015).

On this continent, in recent decades, the human population in most countries has increased, resulting in a drastic decrease in habitat available to large mammal species. This also has led to increases in human-wildlife conflicts in many parts of Africa. Similar to cases in North America and Europe, traditional wildlife management in Africa has been implemented solely based on biological and ecological information, ignoring all human dimensions and social aspects for concern or interest (Natuurnav 1989). However, there are efforts being made currently to achieve coexistence between wildlife and humans. For example, the Communal Areas Management Programme for Indigenous Resources implemented in several African nations utilizes wildlife-based tourism to attract visitors from all over the world, while revenues are used for community development (Child 1996; Gandiwa et al. 2013).

Researches regarding the human dimensions of wildlife management have been implemented throughout Africa and can be found in journals like *Human Dimensions of Wildlife* or others. There have been considerable amounts of research conducted, for example, regarding human-cheetah conflicts in the past few decades that have contributed to integrating human needs into various conservation and communication programs for cheetah management (Marker and Boast 2015). It should be also noted that the international congress of human dimensions of wildlife was held in Africa in January 2016 and was regarded as an important opportunity for stakeholders in Africa to discuss concerns of wildlife management under the human dimensions theme. Meanwhile, there have been no efforts to summarize HDW studies or integrate HDW into African wildlife education, as has been seen in Europe, as far as the author knows (Glikman and Frank 2011, Johansson et al. 2016). While lack of formal higher education programs (under HDW theme) throughout much of Africa is likely responsible for that result, it appears that, perhaps with help of researchers from North America and Europe, the field of HDW and its practices are being recognized in Africa as a useful discipline for decision-making regarding human-wildlife interactions in a region of that world that may need it the most.

2.2.3 *Asia*

Asia is known for having incredibly high biodiversity, which is likely due to the fact that one-third of the world's tropical rain forests (where the world's richest biodiversity is concentrated) are located in Asian countries. The world's most popular wildlife species, known widely as symbols for conservation efforts, are found in Asia and include tigers (*Panthera tigris*), giant pandas (*Ailuropoda melanoleuca*), and orangutans (*Pongo borneo*). Meanwhile, as on other continents, human-wildlife conflicts occur regularly in Asian nations, where herbivores threaten crops and carnivores threaten livestock and people. This fact implies that there is great potential for studying human-wildlife conflicts and how they could be resolved in Asian nations. However, the field of HDW is not well known in many Asian countries and is evidenced by the fact that the number of Asian-based academic papers published in the journal *Human Dimensions of Wildlife* is far fewer than those from Europe or Africa. Meanwhile, efforts to integrate previous human dimensions studies conducted in the regions and/or to have researchers to suggest potential future research needs have not been implemented in Asian countries or by Asian researchers.

Considering all the above, this does not mean that there have been no social studies regarding interactions between humans and wildlife in Asia. In the book by Knight (2013), cultural perspectives of the relationships between local people and wildlife in Asia are well reviewed and edited, mostly utilizing anthropological field research methods. This book explains, for example, (1) how local residents in Bhutan are confronting and dealing wildlife pressures, while maintaining a focus on their traditional lifestyles (Seeland 2013), (2) how local residents perceive and hunt tigers in Indonesia and Malaysia (Boomgaard 2013), and (3) how people in India historically have confronted tigers and wolves (Rangarajan 2013).

Meanwhile, social studies that have utilized social-psychological theories and quantitative approaches to understand local residents' attitudes toward wildlife, or applications of the results therein for decision-making policies, are limited in Asia.

References

Boomgaard P. "Primitive" tiger hunters in Indonesia and Malaysia, 1800–1950. In: Knight J, editor. Wildlife in Asia: cultural perspectives. New York: Routledge; 2013. p. 185–206.

Chapron G, et al. Recovery of large carnivores in Europe's modern human-dominated landscapes. Science. 2014;366(6216):1517–9.

Child B. The practice and principles of community-based wildlife management in Zimbabwe: the CAMPFIRE programme. Biodivers Conserv. 1996;5:369–98.

Decker DJ, Chase LC. Human dimensions of living with wildlife: a management challenge for the 21st century. Wildl Soc Bull. 1997;25:788–95.

Decker DJ, Brown TL, Siemer WF. Human dimensions of wildlife management in North America. Bethesda: The Wildlife Society; 2001.

Decker DJ, Riley SJ, Siemer WF. Human dimensions of wildlife management. Baltimore: The Johns Hopkins University Press; 2012.

Dressel S, Sandstrom C, Ericsson G. A meta-analysis of studies on attitudes toward bears and wolves across Europe 1976–2012. Conserv Biol. 2014;29(2):565–74.

Fazio JR, Gilbert DL. Public relations and communications for natural resource managers. 2nd ed. Dubuque: Kendal/Hunt; 1986.

Gandiwa E, Heitonig IMA, Lokhorst AM, Prins HHT, Leeuwis C. CAMPFIRE and human-wildlife conflicts in local communities bordering Northern Gonarezhou National Park, Zimbabwe. Ecol Soc. 2013;18(4):7.

Glikman JA, Frank B. Human dimensions of wildlife in Europe: the Italian way. Hum Dimens Wildl. 2011;16:368–77.
Gray PA, Boxall P, Reid R, Filion FL, DuWors E, Jacquemot A, Bouchard P, Bath A. Proceedings of the International Union Game Biologists XXI Congress; 1993. p. 151–157.
Jacobson S. Communication skills for conservation professionals. 2nd ed. Washington, DC: Island Press; 2009.
Johansson M, Dressel S, Kvastegard E, Ericsson G, Fischer A, Kaltenborn BP, Vaske JJ, Sandstrom C. Hum Dimens Wildl. 2016;21(2):158–68.
Knight J. Wildlife in Asia: cultural perspectives. New York: Routledge; 2013.
Leong KM, Haigh JT. Deer, people and parks: human dimensions of deer issues in national parks. 2007. http://www.dnr.cornell.edu/research/deerinparks/index.asp. Accessed 12 Dec 2009.
Manfredo MJ, Vaske JJ. Human dimensions of wildlife: introduction. Hum Dimens Wildl. 1996;1:5–6.
Manfredo MJ. Who cares about wildlife? Social science concepts for exploring human-wildlife relationships and conservation issues. New York: Springer; 2009.
Marker LL, Boast LK. Human-wildlife conflict 10 years later: lessons learned and their application to Cheetah conservation. Hum Dimens Wildl. 2015:1–8.
Natuurnav T. The imperial lion. Human dimensions of wildlife management in Central Africa. S Afr J Wildl Res. 1989;19(3):126–7.
Organ JF, Batcheller GR. Chapter 12: Reviving the Public Trust Doctrine as a foundation for wildlife management in North America. In: Manfredo MJ, Vaske JJ, Brown PJ, Decker DJ, Duke EA, editors. Wildlife and society: the science of human dimensions. Washington, DC: Island Press; 2009. p. 161–71.
Osborn FV, Hill CM. Techniques to reduce crop loss: human and technical dimension in Africa. In: Woodroffe R, Thirgood S, Rabinowitz A, editors. People and wildlife: conflict or coexistence. Cambridge: Cambridge University Press; 2005. p. 72–85.
Rangarajan M. The Raj and the natural world: the war against "dangerous beasts" in Colonial India. In: Knight J, editor. Wildlife in Asia: cultural perspectives. New York: Routledge; 2013. p. 207–32.
Seeland K. Wildlife depredations in Jigme Dorji National Park, Bhutan. In: Knight J, editor. Wildlife in Asia: cultural perspectives. New York: Routledge; 2013. p. 131–46.
Thirgood S, Woodroffe R, Rabinowitz A. The impact of human-wildlife conflict on human lives and livelihoods. In: Woodroffe R, Thirgood S, Rabinowitz A, editors. People and wildlife: conflict or coexistence? Cambridge: Cambridge University Press; 2005. p. 13–26.
Vaske JJ, Shelby LB, Manfredo MJ. Bibliometric reflections on the first decade of human dimensions of wildlife. Hum Dimens Wildl. 2006;11:79–87.
Woodroffe R, Thirgood S, Rabinowitz A. The impact of human-wildlife conflict on natural systems. In: Woodroffe R, Thirgood S, Rabinowitz A, editors. People and Wildlife: conflict or coexistence? Cambridge: Cambridge University Press; 2005. p. 1–12.

Chapter 3
Wildlife Management in Japan

Abstract Japan is known for having high biodiversity, with more than 90,000 confirmed animal species, and is listed as one of the 34 "biodiversity hotspots" of the world. From the end of the 1800s to the 1970s, wildlife management focused primarily on "conservation," owing to the fact that most of the major wildlife species were overhunted during that period in Japan. However, thanks to wildlife conservation laws and efforts, species such as wild boar and deer have increased their populations rapidly and have expanded their ranges broadly. Research regarding wildlife management traditionally has been considered as a field of natural science in Japan. Social studies regarding wildlife management conducted in Japan have three primary features: (1) most of these social studies have been conducted because of increasing damage caused by wildlife, (2) a number of these studies have not necessarily followed or utilized findings from human dimensions studies in other parts of the world, and (3) and most of the studies have been published only in Japanese, which has made it nearly impossible for others outside of Japan to understand the social aspects of any studies regarding wildlife management in this country.

3.1 Wildlife in Japan

Located in East Asia, Japan is an archipelago with a total land size of around 38 million hectares and is not connected to any other continent. Although the size of the country is not so large, it is known to have incredibly high biodiversity, with more than 90,000 confirmed animal species (Ministry of the Environment 2008). Japan is listed as one of the 34 "biodiversity hotspots" of the world, particularly rich in endemic plant species that are in danger of extinction (Miller and Spoolman 2015). As of 2010, 241 mammal species, 700 bird species, 97 reptile species, and 64 amphibian species were confirmed in Japan (The Nature Conservation Society of Japan 2010). These numbers are considered rather large, as comparisons with other countries of similar land size and latitude, such as England, reveal only a fraction of the biodiversity, where 64 mammal species, 550 bird species, 7 reptile species, and 7 amphibian species are found (Natural England 2008). It should be noted that a high proportion of endemic species is also characteristic of the Japanese

R. Sakurai, *Human Dimensions of Wildlife Management in Japan*, Ecological Research Monographs, https://doi.org/10.1007/978-981-13-6332-0_3

archipelago, where land area is isolated and small. Among the species mentioned above, ~40% of the terrestrial mammals, ~60% of the reptiles, and ~80% of the amphibian species are endemic to Japan (Ministry of the Environment 2008). Examples of these endemic species include Japanese macaques (*Macaca fuscata*), Japanese serows (*Capricornis crispus*), Japanese giant flying squirrels (*Petaurista leucogenys*), and the Amami rabbit (*Pentalagus furnessi*). There are also large terrestrial mammalian species inhabiting in Japan, including brown bears (*Ursus arctos*), Asiatic black bears (*Ursus thibetanus*), sika deer (*Cervus Nippon*), and wild boar (*Sus scrofa*).

Possessing such great diversity could be explained by the shape of the country. The Japanese archipelago is long, stretching from northerly latitudes with cold temperate forests to more southerly latitudes with warm subtropical forests. Meanwhile, the country also has mountains with altitudes greater than 3000 m. As a result of the diversity of landscapes and temperatures in the Japanese archipelago, there are great numbers of species that can thrive in large ranges of latitudes and altitudes (Ministry of the Environment 2008; The Nature Conservation Society of Japan 2010).

3.2 A Brief History of Wildlife Management in Japan

The history of human-wildlife interactions could be explained as a long-lasting war between people and wildlife (Kaji et al. 2013, p. 20). People have tried many diverse measures to prevent damages occurring by wildlife. As far back as the 1600s, or the Edo period, people were making *Shishigaki* (*shishi* means wild boar, and *gaki* means fences in Japanese) with walls of stones or sand throughout Japan (from Okinawa to the Kanto region). To prevent wildlife invasions into human settlements and agricultural fields, some of the *Shishigaki* stretched as long as 120 km (Kawai and Hayashi 2009, p. 16). Around this time, agricultural damage by wild boar could lead to famine in certain regions, explaining exactly how serious the wildlife problems had become. From around the 1870s, or the Meiji period, people began to expand their agricultural fields and increase their use of wooded areas, which then decreased wildlife habitats. Meanwhile, overhunting of deer and wild boar led to the extinction of these species in some regions of the country. Records from this period show that the government collected 1,500,000 guns that were lent to farmers to prevent wildlife damages. These numbers tell us that farmers and local residents were equipped with more than eight times the total number of guns currently held by hunters in Japan today (The Nature Conservation Society of Japan 2010; Kaji et al. 2013).

The content of hunting laws has changed in the recent decades as human-wildlife relationships and natural resource management also have evolved. Regulations for wildlife hunting in Japan have been issued as long ago as 1873, including specified hunting periods (from December to the end of March) and no-hunting areas (e.g., human settlements); however, since all species were being hunted, the Hokkaido regions saw a rapid decline in the Japanese crane (*Grus japonensis*), as well as the extinction of the Ezo Wolf (*Canis lupus hattai*) (Kaji et al. 2013; Tokida 2015).

In 1892, the regulation was revised to become the "Hunting Regulation," and certain species (e.g., cranes, deer younger than 1 year old) were specified for conservation. Based on the "Hunting Regulation," the new "Hunting Law" was designated and issued in 1895. This law then was revised in 1918 and now is believed to be the basis for the current wildlife conservation law that bans the capturing of certain wildlife species and clarifies that hunting is available only for species designated within the regulation. However, declines in certain wildlife populations continued as the number of hunters increased, and in 1963, the "Hunting Law" was largely revised to become the current "Law of Wildlife Conservation and Hunting." The first article of this law declares the importance of the "conservation of wildlife." The "Law of Wildlife Conservation and Hunting" has ruled that, for example, governors of each prefecture should form a wildlife conservation plan and that hunting licenses are valid only within each prefecture. This law also has designated "wildlife conservation zones" and "no-hunting periods" to limit hunting (Kaji et al. 2013; Tokida 2015).

3.3 Current Wildlife Management in Japan

From the end of the 1800s (the Meiji period) to the 1970s, when Japan was experiencing a period of rapid economic growth, wildlife management primarily was based on "conservation," most likely due to the large amounts of overhunting that were occurring during that period. However, the situation changed dramatically after the 1970s. Thanks to the new wildlife conservation law and various conservation efforts that were being made, populations of species, such as wild boar and deer, began to increase rapidly, expanding their ranges. Urbanization fostered a migration of people from rural areas into more urban areas. The number of hunters decreased rapidly, and wildlife faced fewer pressures from hunting. A rapid increase in certain species led to increases in human-wildlife conflicts, such as agriculture and forest damage, as well as occasional casualties (Kawai and Hayashi 2009).

One of the biggest issues that Japan is facing currently is depopulation and aging. The total population of Japan has been decreasing and is estimated to be at less than 100 million by the year 2050, compared to 127 million in 2013 (Cabinet Office Japan 2014). Globally, Japan has one of the highest rates of depopulation and aging (Miller and Spoolman 2015). It is estimated that by 2060, 1 out of every 2.5 people will be older than 65 years (Cabinet Office Japan 2014). This trend is even more startling in rural areas, such as mountainous regions.

Historically in Japan, people have sustainably utilized resources, such as woods and meadows for fuel and fertilizers, and successional woodlands and grassland around human settlements. This mosaic of ecosystems, including paddy fields, farmlands, wetlands, and human settlements, are called *Satoyama* (*Sato* means villages, and *Yama* means uplands in Japanese) (Takeuchi 2010). *Satoyama* have served as buffer zones between human habitats and wildlife habitats (Kohira et al. 2006; Kawai and Hayashi 2009).

As the so-called fuel revolution, or energy revolution, occurred, woods and coals were replaced by oil, gas, and electricity in the 1950s (Biodiversity Policy Research Group 2002; Kawai and Hayashi 2009). While people traditionally have utilized and managed *Satoyama* forests, there became no need for the *Satoyama* anymore, and these forests were left as unused lands. In the 1970s, Japan experienced a rapid economic growth that changed the economic system of the country. As the country became industrialized and more foods were imported from other countries, the self-sufficiency ratio for food dropped to around 40% (Kawai and Hayashi 2009). As the wood of the *Satoyama* was no longer used, and food was being imported, many people migrated to urban areas to find the jobs that were lacking in rural areas. This fostered urbanization of the country, and the number of farmers rapidly decreased from about six million households in 1960 to about 2.8 million households in 2005 (Kawai and Hayashi 2009). Meanwhile, imports of cheap wood from outside the country began, including from tropical countries with rainforests, and forestry in Japan stagnated; large swaths of forests were left without management. All these factors led to the expansion of wildlife habitats, and an increase in the number of medium- to large-sized mammals, including deer, wild boar, and macaques (Biodiversity Policy Research Group 2002; Tsunoda and Enari 2012). As for deer, habitats have increased 2.5 times, whereas for wild boar habitats have increased 1.7 times since the 1970s (Ministry of the Environment 2015). It is believed that the increase in the number of deer has been ~10 times, from 300,000 in the early 1990s to 3,000,000 currently (Asahi Shimbun 2017). The population density of deer has increased to more than 20 deer per square kilometer in some regions of the country (e.g., Maeji et al. 1999; Nagata et al. 2006). Increases in wildlife species have led to increases in human-wildlife conflicts. Annual costs of agricultural damage caused by wildlife have been estimated at about 20 billion yen (about USD 200 million) from 2008 to 2014 (Ministry of Agriculture, Forestry, and Fisheries 2014). Approximately 19.1 billion yen in agricultural damage was caused by a wildlife in 2014: 6.5 billion yen was due to deer, 5.5 billion yen was caused by wild boar, 1.3 billion yen was the result of damage by macaques, and the rest of damage was caused by other species (Ministry of Agriculture, Forestry, and Fisheries 2014).

In addition, a decrease in the hunter population in Japan also has led to increases in wildlife populations, as their mortality rates have decreased (Tsunoda and Enari 2012). While the country once had more than 500,000 hunters in the 1970s, this number had decreased to less than 150,000 by 2009, resulting from strict gun control policies, as well as a lack of interest or motivation toward hunting among the younger generations (Ueda et al. 2010; Kaji et al. 2013).

Faced with these problems, the "Law of Wildlife Conservation and Hunting" was revised in 1999 so that each prefectural government could make an original conservation and management plan to manage specified wildlife species (such as deer and wild boar) that are rapidly increasing and causing agricultural damage and ecosystem disruption. The plan also targeted those species that are rapidly decreasing and in need of conservation assistance (such as a certain local population of Japanese serow [*Capricornis crispus*]). The idea of adaptive management was introduced in the "Specified Wildlife Management Plan," and each prefectural government was required to implement a long-term wildlife management plan based on scientific data utilizing the PDCA (plan, do, check, and act) cycle. For example, the deer population increased rapidly because of overprotection, as the hunting of female deer was restricted in 1948, and the number of deer that one person could take per day was reduced to one in 1978. As agricultural damage increased and became a serious issue countrywide, female deer were listed once again as a permissibly hunted species in 1994. Since the "Specified Wildlife Management and Conservation Law" was issued in 1999, many prefectural governments started to scientifically monitor deer numbers, setting a goal for reducing the population and increasing the number of deer that hunters could acquire. However, it should be noted that most of the prefectural governments have failed to decrease the deer populations and other wildlife species because the management plans depend entirely on voluntary hunters who are primarily hobby hunters, not subsistence hunters, or because they respond to local government requests to help control nuisance species (Kaji et al. 2013; Tokida 2015).

In 2008, the "Specified Act for Preventing Wildlife Damage" was enforced by the Ministry of Agriculture, Forestry, and Fisheries, although, traditionally in Japan, wildlife conservation and management laws were initiated by the Ministry of the Environment. Through this policy, the national government provided financial support for governments in cities, towns, and villages for implementing interventions to prevent wildlife damage (Kaji et al. 2013; Tokida 2015).

In 2014, the "Wildlife Protection and Proper Hunting Act" was amended. The revision included establishing a certification system for professional culling so that the public sector could engage in wildlife damage prevention. This allowed for capturing wildlife that appeared in residential areas with a specific permission from the governor and lowered the age for acquiring a hunting license from 20 to 18 (Tokida 2015).

3.4 Human Dimensions of Wildlife Management in Japan

Research regarding wildlife management was traditionally considered as a field of natural science and was dominated by studies such as ecology, biology, and veterinary medicine in Japan (Sakamoto 2002; Igota et al. 2011). However, this does not mean that there have been no human dimensions studies conducted in Japan.

One Japanese academic journal that has been publishing various social studies regarding wildlife management is the "Wildlife and Human Society" (*Yaseiseibutsu to shakai*), originally named "Wildlife Conservation Japan" (*Yaseiseibutsu hogo*). Since its establishment in 1995, this journal has published a decent number of papers regarding the social dimensions of wildlife management. These papers include studies regarding the perceptions of local residents and/or visitors toward certain species (e.g., deer, wild boar, macaques), and toward wildlife damage, or regarding policy studies on hunting and/or wildlife management systems. Many of these papers have been written by researchers at universities, whereas some have been written by government officials.

One of the oldest papers that first mentioned and introduced the idea of human dimensions of wildlife management (HDW) in Japan was published in 1996 by Watanabe and Ogura. This paper introduced the USA and Canada as "advanced countries" in wildlife management and briefly explained the development of HDW in the USA in the 1980s. Watanabe and Ogura (1996) utilized the method of HDW and explored local residents' value orientations toward wildlife. They also inquired about preferences for policies in regions where agricultural damage by wildlife was increasing. The result showed that, based on 250 surveys returned by residents, people's recognition of the ecological value of wildlife was connected to the psychological value they held toward wildlife and that older generations had less understanding of ecological values, preferring to get rid of those wildlife species. Watanabe and Ogura (1996) suggested the importance of conducting social studies as well as ecological studies in Japan and also cautioned that such a survey, designed without much consideration of the theoretical background of social science, could decrease the reliability of social studies in the field of wildlife management.

One of the first and only documents to formally introduce and explain the idea, theories, and history of HDW in Japanese was written by Sakurai and Enari (2010). In addition to describing the advantages and effectiveness of the HDW approach, this paper also discusses the challenges that the field faces in the USA, such as lack of studies to actually evaluate how HDW findings are applied and used with regard to wildlife management policies (Sakurai and Enari 2010).

Sociological and qualitative studies associated with wildlife management have been conducted in Japan. Maruyama (2006) looked at the relationship between humans and macaques, addressing the differences in the views of "conservation," where urban residents or conservationists solely see a positive value in having wildlife, whereas local residents believe that there are various meanings to the presence of certain wildlife; for macaques these meanings can be both positive or negative. Maruyama (2006) specifically looked at a case study in the Northern Japan (Aomori), where he qualitatively analyzed how local residents viewed the macaques as "neighbors," and they held various beliefs as to the value of the species. On the other hand, Maruyama (2006) also discussed that various values and meanings of the macaques held by the local residents in this study were not shared outside of the region. In any case, all wildlife management practices enacted in this region would need to consider those local views, and HDW research would be incredibly beneficial in determining how best to serve both the human and macaques populations.

Kikuchi (2006) conducted interviews of more than 200 local residents in Toyooka City, western Japan, revealing how local residents lived, historically, with the oriental white stork (*Ciconia boyciana*). This research also revealed how outsiders and local residents perceived the stork differently. Outsiders considered the stork as a once extinct species that needed conservation, and its reintroduction to the area was a symbol of success [the oriental white stork was once extinct in Japan and was reintroduced from Russia in 2005]; however, local residents considered the storks to be their neighbors that sometimes caused agricultural damage. Kikuchi (2006) also mentioned that the local residents called the storks "cranes," although scientifically speaking, the oriental white stork is not a "crane."

Looking at social studies conducted in Japan regarding wildlife management, there are three features that are worth mentioning. Firstly, most of these studies have been conducted because of increasing agricultural and forest damage, as well as casualties caused by wildlife. As explained in the beginning of this chapter, increases in certain wildlife species populations often lead to increases in human-wildlife conflicts and recently have been one of the biggest social issues in Japan. Many social studies, such as surveys of local residents, have been conducted to better understand their perceptions and experiences with the damages that were incurred and to inquire as to how they believe the issues should be solved, including their demands to the government. Wildlife management in Japan is comprised of three fields: population management, habitat management, and damage management (Muroyama 2009); damage management is considered the main part of wildlife management in Japan. While it is important to utilize social studies for mitigating the agricultural damage by wildlife, it is certainly not the only scope of HDW studies. This trend and these ways of thinking may have led to the belief that many social studies in Japan are only about understanding residents' attitudes for decreasing damage, and not about visitors' impressions of the wildlife species in national parks, or public opinions on invasive species. Although there are such studies regarding public opinion about exotic species or visitors' attitudes toward wildlife viewing, the numbers of these studies are far fewer than studies related to damage prevention.

Secondly, it could be said that a reasonable number of social science studies conducted in Japan regarding wildlife management have not necessarily followed or utilized findings of other HDW studies around the world. Researchers in Japan have a tendency to design studies to reveal the phenomena or factors that the authors are seeking, and not necessarily to test the scientific theories and models (such as social-psychological theories) that previous human dimensions studies have conceptualized and compiled (Aiko et al. 2016). The reasons for this could be related to the previously described concern, where many of these studies are being conducted under the theme of damage prevention. Perhaps, to prevent damage, more applied studies have been favored, rather than those that would fundamentally test certain social-psychological theories; however, this has likely resulted in the lack of studies in Japan that actually test the effectiveness of the HDW approach and theories developed in North America. This also has resulted in a lack of conceptualized theories or methods based on specific Japanese landscapes and cultures, and the current situation in which social science studies of wildlife management still is not recognized as an academic discipline in the country. For the field to be recognized as an academic discipline, methods and theories must be conceptualized so that other studies can follow to test those findings that then will contribute to the theorization of new models. Therefore, even though social studies regarding wildlife management have been conducted for more than 20 years in Japan, it is still unclear how well HDW theories and models (such as the idea of "wildlife acceptance capacity") could work in Japan.

Thirdly, a majority of social studies conducted in Japan are published only in Japanese, which has made it nearly impossible for researchers outside of Japan to understand what had been studied regarding the social aspects of wildlife management in this country. Again, this could be because the goals of many previous studies were only geared toward preventing damage by wildlife in Japan; therefore, authors felt they only needed to publish in Japanese since the expected readers were government officials, local researchers, and other stakeholders in Japan. In addition, since many studies were solely focused on solving local issues, there was likely no reason for authors to look at or test theories from previous studies conducted and published outside of the country.

To consider founding human dimensions of wildlife management as an academic discipline in Japan, newly proposed studies must be based on the results of previous studies, allowing for testing theories and conceptualizing new models. These are the fundamental processes for scientific development. Only by performing studies that have been designed based on previous studies from other parts of the world, and by utilizing, testing, and conceptualizing new hypotheses to adapt to local culture and landscape, can we progress the social studies of wildlife management toward the discipline of HDW. Also by doing so, we can better understand and make suggestions for stronger uses and applicability of the HDW discipline.

3.5 Four Researchers Who Have Contributed to the Development of HDW in Japan

The final section of this chapter closes by introducing some of the Japanese researchers whom, I believe, have contributed most to the development of human dimensions of wildlife management in Japan. It should be noted that these four researchers, who are all members of the Association of Wildlife and Human Society of Japan, have been subjectively chosen here, so there are, of course, many other researchers who have been involved in or have been conducting human dimensions studies in Japan.

Dr. Gouhei Ueda studied the trends of hunters and their perceptions on wildlife management in Japan, publishing several papers on the subject (including Ueda et al. 2004, 2010, 2012). As a government official in Hyogo Prefecture (when these studies were conducted) and as a hunter himself, he was able to reveal the factors that affect Japanese hunters' willingness to hunt using psychological scales.

Dr. Yamabata Naoto conducted studies to understand local residents' attitudes toward macaques and the damage they caused, as well as residential behaviors toward preventing damage (Yamabata 2010a, b, 2011; Yamabata et al. 2015). As a researcher in a government agricultural agency in Mie Prefecture (when these studies were conducted), his continuous work revealed factors that affected residents' willingness to engage in collaborative damage prevention behaviors, as well as the effectiveness of these interventions for actually reducing damage.

Dr. Katsuya Suzuki conducted sociological studies to understand the mechanisms for how local residents and farmers perceive wildlife and damage caused by wildlife, as well as to understand how they interact with wildlife in their daily lives (Suzuki 2007, 2008). He is working currently to revitalize local communities by helping to solve wildlife issues with residents.

Finally, Dr. Takahiro Kubo is a researcher at the National Institute for Environmental Studies (NIES), and he has conducted studies to better understand relationships between local residents/visitors and brown bears in Hokkaido using an environmental economic approach (such as WTP [willingness to pay] for bear conservation action) (Kubo and Shoji 2012, 2014, 2016; Kubo 2014). His research has utilized theories and methods that were developed outside of Japan and has provided perspectives for the link between those theories previously tested on an international scale and local applications for the specific landscapes and society of Japan.

References

Aiko T, Shoji Y, Kuriyama K. Questionnaire survey regarding nature conservation and utilization: handbook of social survey regarding park management, wildlife, and tourism. Tokyo: Kikuchi Shokan; 2016. p. 314. (in Japanese)

Asahi Shimbun. Conflict with wildlife; 2017. February 4, p. 10.

Biodiversity Policy Research Group. 100 key words of biodiversity. Tokyo: Chuohoki; 2002. (in Japanese)

Cabinet Office Japan. Annual report on the aging society. Tokyo, Japan; 2014. (in Japanese). http://www8.cao.go.jp/kourei/english/annualreport/2014/2014pdf_e.html Accessed 12 July 2015.

Igota H, Ueda G, Suzuki M, Yamamoto T, Yoshida T. Wildlife and society: the science of human dimensions. Tokyo: Buneidoshuppan; 2011. (in Japanese)

Kaji K, Igota M, Suzuki M. Science of hunting: for wildlife management in Japan. Tokyo: Asakura; 2013. (in Japanese)

Kawai M, Hayashi Y. Rebellion of wildlife. Tokyo: PHP Science World Research Center; 2009. (in Japanese)

Kikuchi N. Revived oriental white stork. Tokyo: Tokyo University; 2006. (in Japanese).

Kohira M, Okada H, Yamanaka M. 6. Controlled exposure: demographic trends, dispersal patterns, and management of brown bear in Shiretoko National Park. In. McCullough DR, Kaji K, Yamanaka M, editors. Wildlife in Shiretoko and Yellowstone National Parks: lessons in wildlife conservation from two World Heritage sites. Hokkaido: Shiretoko Nature Foundation; 2006. p. 238–41.

Kubo T. Heterogeneous preferences for wildlife management: a latent class approach. Wildl Hum Soc. 2014;1(2):49–60. (in Japanese with English abstract)

Kubo T, Shoji Y. Using choice experiment to assess potential demand for a bear-watching tour: a case study in Daisetsuzan National Park, Japan. Wildl Conserv Jpn. 2012;13(2):9–18. (in Japanese with English abstract)

Kubo T, Shoji Y. Spatial tradeoffs between residents' preferences for brown bear conservation and the mitigation of human-bear conflicts. Biol Conserv. 2014;176:126–32.

Kubo T, Shoji Y. Public segmentation based on the risk perception of brown bear attacks and management preferences. Eur J Wildl Res. 2016;62(2):203–10.

Maeji I, Yokoyama S, Shibata E. Population density and rage use of sika deer, *Cervus nippon*, on Mt. Ohdaigahara, central Japan. J For Res. 1999;4:235–9.

Maruyama Y. Environmental issues of macaques and people: between nature conservation and damage regarding Japanese macaques. Kyoto: Showado; 2006. (in Japanese).

Miller GT, Spoolman SE. Living in the environment. 18th ed. Stamford: Cengage Learning; 2015.

Ministry of Agriculture, Forestry and Fisheries. Change of agricultural damage by wildlife. 2014. (in Japanese). http://www.maff.go.jp/j/seisan/tyozyu/higai/h_zyokyo2/h26/pdf/160122-b.pdf. Accessed 11 Jan 2017.

Ministry of the Environment. The wildlife in Japan. Tokyo: Wildlife Division, Nature Conservation Bureau; 2008.

Ministry of the Environment. Results of survey of population and habitats of deer and boars for implementing specified wildlife capturing program based on revision of wildlife law. 2015. (in Japanese). http://www.env.go.jp/press/100922.html. Accessed 11 Jan 2017.

Muroyama Y. Chapter 3: Wildlife management. In: Kawai M, Hayashi Y, editors. Rebellion of wildlife. Tokyo: PHP Science World Research Center; 2009. p. 55–78. (in Japanese).

Nagata K, Koboyashi T, Yamane M, Tamura A, Kuribayashi H, Takii A. Report of the sika deer (*Cervus Nippon*) population research for sika deer management. Bull Kanagawa Prefecture Nat Environ Conserv Center. 2006;3:28–36. (in Japanese).

Natural England. Chapter 3 Biodiversity. In: State of the natural environment 2008. 2008. http://publications.naturalengland.org.uk/publication/31043. Accessed 29 Sept 2017.

Sakamoto M. Book review of "Wildlife issues" by Hayama, S. Enviorn Pollut (*Kankyo to kogai*). 2002;31:70. (in Japanese).

Sakurai R, Enari H. What is human dimensions? Development of social study approach in wildlife management. Wildl Forum 2010; 14(3-4):16–21. (in Japanese).

Suzuki K. Monkey damage in the Shimokita Peninsula and farmers' complicated recognition of the damage: a dilemma between multiple agriculture and animal control measures. J Environ Sociol. 2007;13:184–93. (in Japanese with English abstract).

Suzuki K. How can we resolve human-wildlife conflicts?: analyses on developing processes of wildlife problems focusing on local people's cognitive structures. J Environ Sociol. 2008;14:55–69. (in Japanese with English abstract).

Takeuchi K. Rebuilding the relationship between people and nature: the Satoyama Initiative. Ecol Res. 2010;25:891–7.

The Nature Conservation Society of Japan. Wildlife conservation and law from ecological perspectives: toward conservation of biological diversity. Tokyo: Kodansha; 2010. (in Japanese).

Tokida K. History of wildlife management in Japan and amendment of the Wildlife Protection and Proper Hunting Act in 2014. Wildl Hum Soc. 2015;3(1):3–11. (in Japanese with English abstract).

Tsunoda H, Enari H. Chapter 7: Ecological role and social significance of reintroducing wolves in Japan under the shrinking society. In: Paula A, Crussi HF, editors. Wolves: biology, behavior and conservation. New York: Nova Science Publishers; 2012. p. 177–98.

Ueda G, Kanzaki N, Kodera Y. Present conditions of hunter activities and attitudes in Shimane Prefecture. Wildl Conserv Jpn. 2004;9(1):9–22. (in Japanese with English abstract)

Ueda G, Kanzaki N, Koganezawa M. Changes in the structure of the Japanese hunter population from 1965 to 2005. Hum Dimens Wildl. 2010;15:16–26.

Ueda G, Kodera Y, Krumada T, Takeuchi M, Sakurai R, Sasaki C. Factors affecting hunting desertion in Japan: proposal for retention measures. Wildl Conserv Jpn. 2012;13(2):47–57.

Watanabe O, Ogura S. Relationship between perceptions toward wildlife values and opinions about wildlife management policy in rural areas of central Japan. Wildl Conserv Jpn. 1996;2(1):1–15. (in Japanese with English abstract)

Yamabata N. Mitigate effect on damage to food crops achieved by collaboration of a whole village for chase-off of monkeys: verification in 6 area of Mie Prefecture. J Rural Plan. 2010a;28:273–8. (in Japanese with English abstract)

Yamabata N. Effect of improved countermeasures to agricultural damage by wildlife on farmers' awareness of farmland management: a study of community in Mie Prefecture. J Rural Plan. 2010b;29:245–50. (in Japanese with English abstract)

Yamabata N. Effect of chasing away by village on the home range and appearances of a macaques group: verification in 7 area of Mie Prefecture. J Rural Plan. 2011;30:381–6. (in Japanese with English abstract)

Yamabata N, Kuki Y, Hoshino S. Did continuous countermeasures against agricultural damage by wildlife effect the settlement's social capital? Case study in a area of Mie Prefecture. J Rural Plan. 2015;34(3):369–75. (in Japanese with English abstract)

Chapter 4
Studies on the Human Dimensions of Black Bear Management in Japan

Abstract Human-bear conflicts are one of the most serious human-wildlife issues in Japan. Every few years, when there is a lack of natural foods, Asiatic black bears (*Ursus thibetanus*) appear in massive numbers near human settlements. In the past, this has resulted in thousands of bears being captured and killed, while hundreds of people have been injured in bear attacks. A series of human dimensions studies were conducted in Japan regarding black bear management. First, a study on content analysis was conducted to identify the characteristics of media coverage in Japan regarding black bear issues. The study revealed that a majority of newspaper articles regarding bears discussed problematic interactions, such as the fear felt by people toward bears. Second, interviews of local residents were conducted at one study site (Tajima region; northern part of Hyogo Prefecture, western Japan) to better understand the residents' perception of bears; the interviews were analyzed qualitatively. While many local residents expressed negative feelings and hatred toward bears, the interviews also revealed other aspects of human-bear interactions. Many people in the community considered the bears as their neighbors in the mountain. Third, surveys of local residents were conducted to better understand not only their perception of the bears but also the degree of actions taken to prevent damage by bears. Based on quantitative analysis, the surveys revealed that although bear-induced damage experienced by two towns in Tajima region was similar, the rate of reporting and taking actions to prevent further damage was significantly lower in one town (Tanto) than in the other (Kami). Finally, research was conducted to evaluate the effects of a community bear education seminar and to identify the potential for, as well as the limitations of, changing peoples' behaviors through educational means. A pre-post survey identified that actions requiring significant manpower were difficult for local residents to adhere to, even though they had intentions to follow through immediately after the seminar.

R. Sakurai, *Human Dimensions of Wildlife Management in Japan*, Ecological Research Monographs, https://doi.org/10.1007/978-981-13-6332-0_4

4.1 Introduction

The Asiatic black bear (*Ursus thibetanus*, hereafter "black bear") is one of the largest wild animals in Japan. It is listed as "vulnerable" in the International Union for the Conservation of Nature Red List of threatened species, and international trade in this bear is controlled by the Convention on International Trade in Endangered Species (Japan Bear Network 2007a, b; Garshlis and Steinnetz 2008). In Japan, black bears inhabit the mainland (Honshu) and Shikoku areas, while they are considered extinct in Kyushu. Estimating the population of black bears is difficult, and there is still debate on the actual numbers of black bears in Japan. Some have estimated the population to be 8400–12,600 (Japan Bear Network 2007a, b). Some researchers believe that previous studies have underestimated the black bear population, stating that it could be much more than 10,000 (Onishi 2008 p. 193).

Traditionally, black bears have been depicted as friendly creatures in folklore, such as in *Kintaro*, in which a black bear serves as a friend to the main character. Bears have been popular in Japanese culture, and people have associated black bears with mountain spirits and gods. In addition, since bears resemble humans in appearance, black bears are sometimes referred to as "mountain man (*yamaotoko*)" or "lonely person (*sabishigariya*)" (Knight 2003).

Black bears mainly eat mast and other foods, including fruiting trees, herbaceous plants, and invertebrates. Since they rely heavily on mast, black bears are believed to travel between forests areas and wander around searching for food during lean years when mast production is low (Yokoyama 2009). This has led to a large number of black bears appearing in human settlements during drought years, which occur once every few years; most recently, droughts have occurred in 2004, 2006, 2010, and 2014. In these years, as massive numbers of black bears appear around human settlements, increased damage to agricultural crops and increased human casualties are seen. As a result, many black bears are captured and killed because the response by local residents in Japan, when black bears are observed, is to ask the local government or police to kill the bears. Police and government, in turn, ask hunters to kill the bears (Huygens et al. 2001). For example, in 2006, a total of 4340 black bears were captured and most were killed, while 142 people were injured and 3 were killed in black bear attacks (Ministry of the Environment 2007a). Similarly, in 2010, a large number of bears appeared around human settlements, and 3358 black bears were captured; 3074 of those were killed (Ministry of the Environment 2016a). During the same year, 142 people were injured and 2 were killed by black bears (Ministry of the Environment 2016b).

Having these appearances of massive numbers of bears every few years that lead to human-bear conflicts, including shocking incidents of casualties, local residents tend to fear black bears and often prefer to destroy the bears causing problems near residential areas (Hugyens et al. 2001; Ministry of the Environment 2007b; Uchikoshi 2007). As previously mentioned, this likely is a very different response from how people traditionally respected bears as mountain spirits.

To conserve and sustainably manage black bear populations in Japan, management agencies must garner public support toward management policies and help the public understand these policies. In addition, to mitigate the types of human-bear conflicts previously mentioned, citizen participation is critical for the implementation of damage prevention actions. For these reasons, conducting human dimensions studies on black bear management procedures is important and provides for a better understanding of public perception, values, and attitudes toward black bears, public willingness to engage in damage prevention behaviors, and other factors affecting those behaviors. However, at this point in time, there have not been many human dimensions studies conducted in Japan regarding black bear management.

This chapter introduces a series of human dimensions studies on black bear management conducted in Japan. The first part of this section describes the results of media coverage content analysis regarding black bear issues in Japan. The second part of this section focuses on a series of studies conducted in the northern part of Hyogo Prefecture (located in western Japan) and includes the results of qualitative studies, such as interviews, as well as quantitative studies, such as local resident surveys regarding stakeholder perceptions of black bears and damage prevention behaviors. The last part of this chapter describes research that was conducted to evaluate the effectiveness of a community seminar regarding black bear issues. The study reveals how participants' perceptions and willingness to take part in damage prevention actions changed after the seminar.

Please keep in mind that this chapter was written based on academic studies and the resulting papers written by Sakurai, R (the author). Section 4.2 was based on Sakurai et al. (2013a), Sect. 4.3 was based on Sakurai et al. (2011), Sect. 4.4 was based on Sakurai et al. (2012), and Sect. 4.5 was based on Sakurai et al. (2013c).

4.2 Content Analysis Study on Media Coverage of Black Bear Issues

4.2.1 Why Media Analysis?

To begin human dimensions studies on black bear management in Japan, a study was conducted on the media coverage of black bear issues and associated management procedures. Media coverage is one of several factors, along with social norms, knowledge, and direct and indirect experiences, that affect peoples' perceptions of and attitudes toward certain topics and issues (Slovic 1987; Detjen 1995; Muter et al. 2009); media coverage has a tremendous impact. Tragic incidents, such as wildlife-related human casualties, likely are covered in the media, and these stories are read or seen by a certain audience. Two types of well-studied and discussed influences that the media have on public opinions are agenda-setting and framing effects (Price et al. 1997; Boykoff and Rajan 2007). Agenda-setting effects occur when media outlets select newsworthy stories to produce articles or reports that affect the audience's opinions about those issues (Price et al. 1997). For example, an audience could interpret an issue as important if it was covered in a long article or if many articles appear on the subject in high frequency. Meanwhile, the manner in which the media portray and frame an issue in their articles could affect how well an audience understands the topic (Price et al. 1997). This is called the framing effect.

As for the current study, I focused on newspaper articles to understand how the Japanese media portray bear issues in Japan. While television and internet have become main media sources globally, newspapers still play an important role in countries like Japan; most people depend on newspaper to gain information. Some of the world's largest newspapers, in terms of circulation, are published and read in Japan: the Yomiuri has a circulation of approximately 24,067,000, the Asahi has a circulation of approximately 12,121,000, and the Mainichi has a circulation of approximately 5,587,000 (World Association of Newspapers 2005). Based on agenda-setting and framing effects, as well as the importance of newspaper as sources of information in Japan, it is expected that newspaper articles influence how people perceive certain issues. A previous study revealed that local residents depend on newspaper articles to learn about bears and bear issues in Japan (Sakurai et al. 2012). For some time, biologists have insisted that mass media negatively affects the public's perceptions of bears, providing context for people to feel the species is dangerous because most media stories are about damage and casualties caused by bears (Japan Bear Network 2007a, b; Yokoyama 2009).

One concept, the social amplification of risk, explains how the mass media have the ability to increase the audience's perception of risk (Jaeger et al. 2001). According to this concept, the media can amplify the audience's perception of risk about a certain event, even though the actual risk is low. The media often has a tendency to report situations as more dangerous than they really are, increasing the audience's perception of risk (McQuail 2005). Risk perception, in turn, can influence the audience's attitudes, beliefs, and behaviors (Gore et al. 2005). Because we know that the media has these effects on audiences, it is important to understand how bears and bear incidents are portrayed in the media in Japan. This information could help managers to tailor communication messages and outreach regarding bears specifically for stakeholders and residents.

4.2.2 Methods

For this study, content analysis was conducted of media coverage on bears. Content analysis systematically evaluates texts and communication artifacts using valid quantitative measurement methods (Riffe et al. 2005). This study specifically focused on newspaper articles, analyzing how bears and bear incidents were portrayed in the papers. Articles were analyzed from years of both high and low bear appearance rates, as well as from both local and national newspapers. The fiscal year 2006 (from April 2006 to March 2007) was selected because a massive number of bears appeared around human settlements during that time, caused by a shortage of bear-preferred foods in forests. As a result, the highest number of black bears on record was captured (n = 4839) and killed (n = 4335), while more than 140 people were injured and 3 were killed in black bear attacks (Ministry of the Environment 2008). The fiscal year 2005 (from April 2005 to March 2006) was also selected for this study, as a relatively low number of bears appeared around human settlements during this year (719 black bears captured; 56 people injured) (Ministry of the Environment 2008).

The following two hypotheses were tested:

1. In the year when high number of bears appeared, more articles (numerically), longer articles, and more articles with increased negativity were published compared with the year when low number of bears appeared.
2. Local newspapers circulated within regions that are more likely at risk of bear appearances published more negative articles with risk explanation than did national papers.

The Yomiuri newspaper, previously mentioned as one of the world's largest newspapers in circulation (World Association of Newspapers 2005), was chosen as the national paper. The Shinano Daily was chosen as the local newspaper for this analysis. It is the most read newspaper in Nagano Prefecture, which is where the highest number of black bears were captured in 2006 (794 black bears) and the highest number of human casualties occurred (16 cases).

In the Yomiuri, most of the articles regarding bears used the Japanese Katakana character "クマ" for bear, while in the Shinano Daily, most of the articles used the Chinese character "熊" for bear. Both characters, "クマ" and "熊," are used to refer to Asiatic black bears and brown bears (*Ursus arctos*), which also inhabit Northern Japan. Using an electronic search engine for the Yomiuri newspaper, 1046 articles were identified from 2005 (from April 2005 to March 2006), and 1462 articles were identified from 2006 (from April 2006 to March 2007) that included the word "クマ" in the text. Similarly, in the Shinano Daily newspaper, 764 articles were identified from 2005 (from April 2005 to March 2006), and 913 articles were identified from 2006 (from April 2006 to March 2007) that included the word "熊" in the text. To extract articles that actually discussed Asiatic black bears, I looked at the headlines from all articles, and those that did not specifically mention the Asiatic black bear were eliminated. This narrowed the samples to 194 articles from 2005 and 624 articles from 2006 in the Yomiuri, and 31 articles from 2005 and 118 articles from 2006 in the Shinano Daily. For the analysis, an adequate sample size (Lombard et al. 2002) of 100 random articles was chosen for 2005 and 2006 from the Yomiuri newspaper; all articles from the Shinano Daily were used.

A coding sheet was developed explaining the criteria and categories for coding articles and was modeled after previous studies from the USA on context analysis of the media coverage of bears (Gore et al. 2005; Siemer et al. 2007). Two trained coders evaluated articles following guidelines, such as "topics discussed," "impacts of bear issues," "types of solutions," "risks," and "sources of information."

To identify reliability between coders, percent agreement was considered for the different articles reviewed; using ten "dummy" bear articles in a different newspaper (the Asahi), percent agreement was calculated as 92.4%. Then, coders reviewed the items that they had coded differently and discussed how they could improve the level of agreement for the code. After coding ten other "dummy" bear articles, the percent agreement increased to 93.0%. Scott's Pi test was conducted for some samples articles ($n = 29$) that had been evaluated by the coders; the resultant Scott's Pi coefficient was 0.74. The calculated percent agreement and the results of Scott's Pi test implied that the reliability of the coders was fairly acceptable. Analysis of variance (ANOVA) and chi-square tests were used to identify differences between compared groups. SPSS version 18 (SPSS Inc. Chicago, USA) was used for the analysis, and $p < 0.05$ was considered statistically significant.

4.2.3 Results

As for article lengths, the mean number of words per article in the national paper (the Yomiuri) was 332 ± *SD* 297 in 2005 and 365 ± *SD* 261 in 2006. The mean number of words per article in the local paper (the Shinano Daily) was 380 ± *SD* 156 in 2005 and 499 ± SD 236 in 2006. Statistical analysis revealed that articles in the local newspaper were significantly longer than those in the national paper ($F = 7.988$, $p < 0.01$) and that articles in 2006 (with high bear appearance) were longer than those in 2005 ($F = 5.545$, $p \leq 0.05$).

As for the coverage of bear-related issues, more articles mentioned the increased likelihood of threats or attacks by bears in 2006 (61%) than in 2005 (36%) ($x^2 = 18.440$, $p < 0.01$). Further, more articles in 2006 mentioned that bear-related issues were increasing than in 2005 ($x^2 = 26.832$, $p < 0.01$). Articles in the local newspaper more often mentioned the probability of threats and attacks by bears ($x^2 = 13.355$, $p < 0.01$), unacceptability of bear-related risks ($x^2 = 11.297$, $p < 0.01$), and increasing rates of bear-related risks ($x^2 = 19.357$, $p < 0.01$).

As for the topics covered in the articles, "problematic interactions with bears," "government response to bear incidents," and "bear sightings" were the most commonly mentioned (>45%), whereas topics regarding "bear conservation" and "education" were covered in a limited number of articles (<20%). More articles regarding problematic interactions with bears were published in 2006 than in 2005 ($F = 14.004$, $p < 0.01$). "Worry or fear of human injuries caused by bear attacks" (51%) and "damage caused by bears to property or humans" (39%) were the impacts most often mentioned in the papers, whereas only a limited number of articles (7%) discussed the ecological aspects of the local bear population, such as maintaining a viable and sustainable bear population (Table 4.1).

The most frequently mentioned solutions for human-bear conflicts were "teach people how to live with bears" (51%) and "increase warnings to people" (46%), whereas a reasonable number of articles mentioned "trap and move bears" (17%), "lethal control of problem bears" (15%), and "capture and release after aversive

Table 4.1 Most frequently mentioned topics in the national and local newspapers

		Frequencies				
		National paper		Local paper		
		2005 ($n = 99$)	2006 ($n = 100$)	2005 ($n = 31$)	2006 ($n = 118$)	Mean
Most frequently mentioned topic	Problematic interactions with bears	49%	57%	32%	69%	59%
	Government response to a bear incident	40%	43%	58%	59%	49%
	Bear sighting	40%	54%	58%	43%	47%
Most frequently mentioned effect and interaction	Worry or fear of human injuries caused by bears	46%	46%	52%	61%	51%
	Damage caused by bears to either property or humans	38%	29%	32%	49%	39%
	Human injuries or deaths caused by bears	35%	24%	23%	42%	33%
Most frequently mentioned solution	Teaching people how to live with bears	51%	45%	61%	54%	51%
	Increasing warnings and inducing caution	49%	43%	52%	44%	46%
	Trapping and moving bears	5%	8%	23%	32%	17%
Most frequently mentioned source of information	Police	46%	45%	19%	10%	31%
	Local governments	19%	14%	45%	33%	25%
	Prefectural governments	23%	19%	10%	25%	22%

Revised from Sakurai et al. (2013a)

conditioning to make bears feel afraid of people" (11%) as solutions. The local paper published significantly more articles regarding "trap and move bears" ($F = 22.751$, $p < 0.01$) and "capture and release bears after aversive conditioning" ($F = 14.201$, $p < 0.01$) than did the national paper, whereas papers in 2006 published more articles regarding "lethal control of problem bears" ($F = 13.782$, $p < 0.01$) and "capture and release bears after aversive conditioning" ($F = 4.886$, $p < 0.01$) than in 2005 (Table 4.1).

While 51% of articles mentioned the probabilities of threats or attacks by bears, only 1% provided supporting statistics to back up those statements. As for sources of information, the police (31%), local governments (25%), prefectural governments (22%), regular citizens (19%), researchers (16%), and hunters (10%) were often referred to in articles. While the national paper was more likely to mention "police" ($F = 32.340$, $p < 0.01$) as a source of information, the local paper more often mentioned "local governments" ($F = 32.340$, $p < 0.01$), "regular citizens" ($F = 6.856$, $p < 0.01$), and "researchers" ($F = 16.655$, $p < 0.01$) (Table 4.1).

4.2.4 Discussion

The study found that a majority of newspaper articles regarding black bears were focused on problematic interactions, such as how people felt fear and worry toward bears. In other words, most articles could be categorized as discussing human-bear conflicts. This trend that wildlife is more likely to appear in the media when they cause problems for people has been studied and identified in previous research (Gore et al. 2005; Siemer et al. 2007).

Several characteristics were revealed in this study. First, most articles mention the necessity of changing human behaviors, such as by teaching people how to live with bears and warning people about damage to help reduce issues. However, in reality, most bears that appeared around human settlements were captured and killed, implying that while media promotes the necessity of change in human behaviors, destroying bears has remained the primary management action. This could be caused by the negative attitudes of local residents as well as a lack of acceptance of bears (Sakurai and Jacobson 2011).

Second, bear-related risks were mentioned more often in 2006, the year in which high number of bears appeared (2006), and in the local newspaper. In addition, only 1% of the articles mentioning the frequencies of risks were supported by statistical analyses. If the majority of bear-related articles were about risks, this could have a targeted effect on readers, perhaps causing negative attitudes toward this species. In a US newspaper, articles regarding wildlife issues are more to likely include supporting statistics, such as "1 out of 5 million encounters with bears generates a fatal attack" (Gore et al. 2005). The fact that a majority of articles regarding bears in Japan were about risks and most contained no supporting statistics is likely to result in an increase in reader concerns regarding this species.

Third, primary information sources for articles regarding bears were, in many cases, the police, not government or experts. This reflects the fact that, in Japan, the police usually handle wildlife issues and communicate with the media. However, since the police are not experts of bear ecology or human-bear conflicts, they are limited in their abilities to provide thorough information about bear characteristics, causes and effects of bear problems, and potential management options.

Even in the current Internet Age, newspapers remain an important source of media particularly in Japan, and this study found that the methods being used by newspapers were negative in nature and very likely to cause increased problems as opposed to helping relieve some of the tension. Government officials and bear ecology experts must reach out to reporters to ensure that articles provide objective and scientific information regarding bears, and it is necessary to include the appropriate statistics for the possibility of bear threats. The role of a newspaper should be to disseminate accurate information from reliable sources, and by including objective, broader, and more detailed information about bears, people can equip themselves with the appropriate knowledge regarding local bear issues and with accurate perceptions of the risks involved.

4.3 Perceptions of Local Residents Regarding Black Bears: Qualitative Approach

4.3.1 Background of the Issue: Human-Bear Conflicts in Hyogo Prefecture

Hyogo Prefecture (Fig. 4.1) is located in western Japan and is inhabited by the "eastern Chugoku population" and the "northern Kinki population" of black bears. These two populations have been categorized as Rank A in the 2003 Hyogo Prefecture Red Data Book (Hyogo Prefecture 2003). This means they are considered endangered or they are populations that require urgent or rigorous conservation measures. In 1992, bear hunting was voluntarily restricted in the prefecture, and then in 1996, it was completely banned. In 2003, the Phase 1 Black Bear Conservation and Management Plan (a 4-year plan) was initiated, followed by the Phase 2 Black Bear Conservation and Management Plan in 2007 (Hyogo Prefecture 2009). Under this policy, bears that repeatedly caused damage were captured for behavior

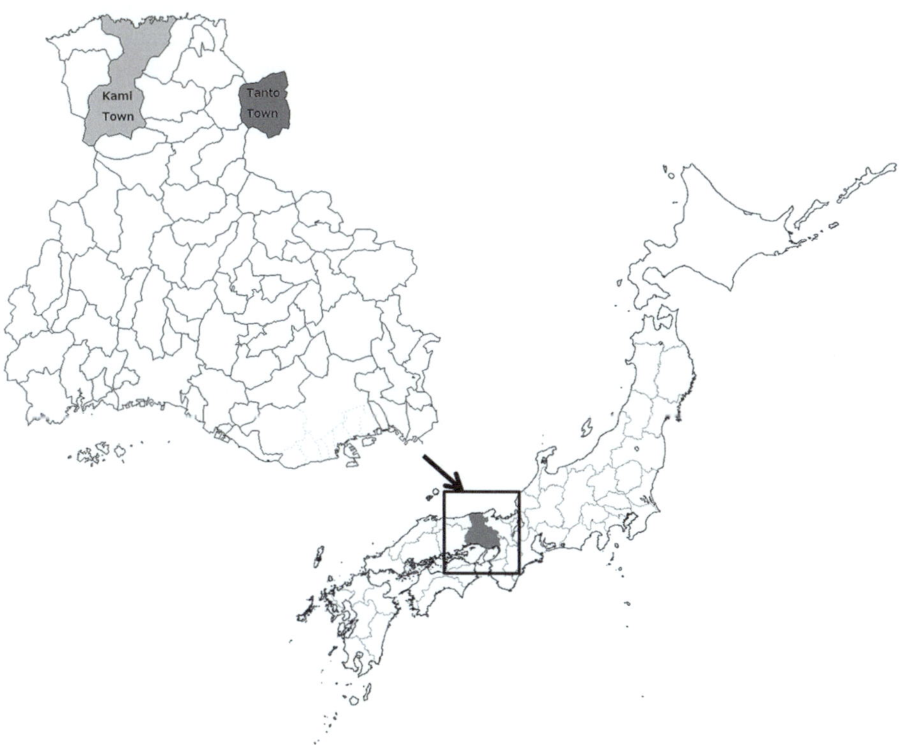

Fig. 4.1 Location of Hyogo Prefecture (shaded area, right map) and an inset map showing Kami town and Tanto town in the prefecture (shaded areas, left map)

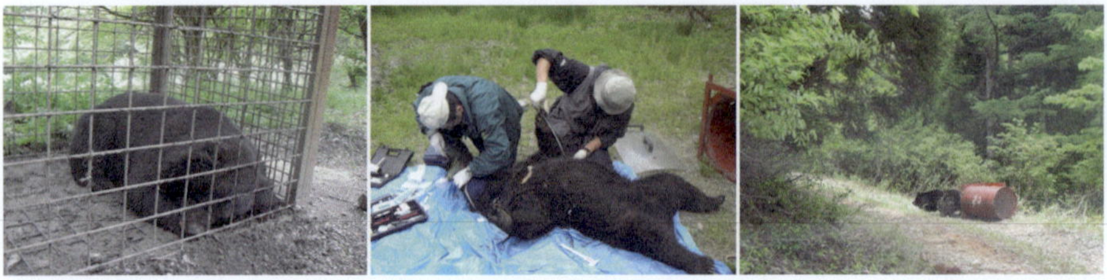

Fig. 4.2 An anesthetized bear captured and sleeping in a trap (left). The same bear being tagged after anesthetization to track its behaviors (middle). The same bear being released in deep forest after aversive conditioning (e.g., scaring the bear with noisy fire crackers; right). (All photos were taken by the author in the study area of Tajima region)

modification and released after aversive conditioning (e.g., using noisy fire crackers to scare the bears; Fig. 4.2). Killing the bears was allowed only if the conditioning was not considered effective or the risk of inflicting human injury was high.

In 2007, the Hyogo Prefectural government established the Wildlife Management Research Center, Hyogo (a research facility that specializes in wildlife), and launched a wildlife specialist program in which local government officials who have the appropriate management skills can take action on site. Hyogo became one of the first prefectures in Japan to work on initiatives to reduce damage caused by bears while avoiding extinction of sensitive populations (Japan Bear Network 2007b). Due to these efforts, the bear population in the prefecture has been increasing (Hyogo Prefecture 2009), but this has also resulted in increased conflicts between local residents and the bears (Yokoyama et al. 2008). In 2011, the bear population was estimated to have increased to approximately 650 individuals (Sakata et al. 2011), causing a status change in the Red Data Book of Hyogo Prefecture from Endangered Species Rank A to Rank B (Hyogo Prefecture 2011). On the other hand, human-bear conflicts have been increasing in these regions (Yokoyama et al. 2008), and in 2006, a year in which a high number of bears appeared, agricultural damage amounted to approximately $180,000 (Hyogo Prefecture 2009). During the period 1998–2009, there were five cases of reported human casualties caused by bears in Kami town, but no casualties in Tanto town (Hyogo Prefecture 2009). In 2010, a lack of food resources in forests resulted in reports of 1623 bear sightings around these towns (Wildlife Management Research Center, Hyogo 2011) and 5 human casualties occurred in Hyogo Prefecture in a single year, the highest on record (Ministry of the Environment 2011). In Hyogo, where steady efforts have been made to ensure the scientific management of bears, it is essential for people to find ways to coexist with bears.

Residents and their communities often unexpectedly encounter bears that are attracted by persimmon trees and improperly managed garbage (Japan Bear Network 2007a). To prevent agricultural damage and human injury, local governments must educate citizens and encourage them to take initiatives to reduce damage while raising bear awareness. For example, persimmon and chestnut trees attract bears and other wildlife, such as macaques, because the fruits are eaten by those species. One

Fig. 4.3 Trees wrapped with corrugated galvanized iron sheets help prevent wildlife from climbing them. (All photos were taken by the author in the study area Tajima region)

management method to help prevent wildlife from being attracted to the trees is to wrap the trunks with corrugated galvanized iron sheets so that the bears cannot climb them (Fig. 4.3.); however, these types of damage prevention activities are not effective unless the entire community is committed to discouraging bear encounters (Yokoyama et al. 2008). Local governments take action by raising awareness in and around bear-frequented areas; therefore, government officials encourage reports of bear sightings from local residents to help prevent bear encounters.

To promote community-based efforts that encourage residents to work together on damage prevention activities, it is essential for residents and stakeholders in their respective communities to share issues and concerns, set targets for solving human-bear conflicts, and build a consensus (Decker et al. 2001). To provide a better understanding of the situation and to help prevent damage, it is necessary to conduct research on both (1) the biological and ecological aspects of bears and (2) the social aspects of including human perceptions of bears and their behavioral intentions (Baruch-Mordo et al. 2009).

People's perceptions of the bear issues at hand can be diverse depending on their positions, experiences, social attributes, etc., which complicates the situation and makes it difficult to build a consensus toward action (Japan Bear Network 2007b, Teel and Manfredo 2010). To reveal public perceptions on a certain topic, various survey methods are available, including observations, questionnaire surveys, and telephone surveys. The first step in a series of studies designed to understand public perceptions of bears was to conduct interviews in each of the communities to better grasp the current situation and provide a "bigger picture" of the problems to be addressed. To understand the relationship between local residents (communities) and bears, it was necessary to analyze their interactions; this was accomplished by documenting the narratives of residents who had experienced bear interactions. Specifically, this study aimed to solicit opinions about bears from residents and stakeholders in the Tajima region (Fig. 4.1), to understand current community statuses, and to reveal which factors cause residents to consider bears a nuisance (e.g., agricultural damage). Interviews of stakeholders were conducted in Kami town (which is inhabited by the eastern Chugoku bear population) and in Tanto town, Toyooka City (which is inhabited by the northern Kinki bear population). In Japan, the smallest administrative division is called a "district," and each district generally has a district chief. Interviews were conducted of 22 stakeholders, including district chiefs, municipal government staff in charge of wildlife damage, and hunters, from May to July 2010 in Kami and Tanto towns. Each interview took approximately 90 min. In addition, interviews were conducted in homes and offices of 55 local residents. In district U, a group discussion was held with six stakeholders. The in-depth interview technique was used to allow people to talk freely, answering questions about specific topics.

4.3.2 Interactions Between Local Residents and Bears in the Northern Part of Hyogo Prefecture

4.3.2.1 Negative Attitudes Toward Bears and Recognition of Changes in Residents' Lifestyles

From interviews, it became clear that perceptions of bears differed depending on the position of the respondent. For example, previous studies have shown that community representatives (e.g., community chief, village chief) often express the opinions of all or some of the residents (Suzuki 2008). For this set of interviews, Mr. A, chief of district Z in Kami town, expressed concerns about the safety of residents when he was asked about bears:

"I am not anxious about bears during the daytime, but residents may encounter bears in the morning and at night. I feel anxious when they go to work at around 6:00 a.m. I try to tell them to take precautions. I am most concerned about children when they go to school. Adults do not go out when the risk is high. When children must go out at night, a family member gives them a ride."

As described below (underlined ⓐ), Mr. B, chief of district Y in Kami town, stated that his role was to express the opinions of the residents:

"Bears are scary. The residents share this feeling. We do not go into the mountains because we are elderly. I urge people who go into the mountains to bring bells. <u>It is my role to properly communicate the residents' feelings to the local governmentⓐ</u>."

Mr. C, chief of district X in Tanto town, expressed a negative opinion about bears. While this statement definitely reflects a personal opinion about bears, his statement also is believed to reflect the opinions of local residents.

"I do not understand why bears are protected. <u>Weⓑ</u>, the local residents, are troubled. We do not have any problems without bears. Why shouldn't bears be allowed to go extinct? <u>Iⓒ</u> believe that animals that cause fear should be eliminated. Bears should be protected at a zoo."

This statement reflects both Mr. C's personal opinion (underlined ⓒ) and the opinions of the community residents ("We" [underlined ⓑ] is considered to refer to the people in the community).

The following statements were made by local residents about bears:

- "Bears are not welcome. I am scaredⓓ. Measures should be taken so that any individual can go into the mountains alone" (a woman from district W in Kami town).
- "Bears come when mast is scarce. We encounter bears because the trees in the mountains have been cut downⓔ" (a woman from district V in Kami town).
- "Bears used to be hunted and killed. Recently, bears are only captured and released, so the population is increasing. Bears should be killedⓓ. We did not encounter bears before. People went into the mountains to collect chestnutsⓕ. We started to encounter bears after tall trees were cut downⓔ" (a woman from district Z in Kami town).
- "Bears come due to a lack of mastⓔ. Trees have started to grow again, but people no longer work in the mountainsⓕ" (a man from district U in Tanto town).

The women from districts W and Z expressed negative feelings toward bears (underlined ⓓ). The local residents stated that they began encountering bears because there is no longer sufficient mast available in the mountains (underlined ⓔ). Some residents are even sympathetic to the bears because they know that the bears have no choice but to enter the communities to look for food. The man from district U in Tanto town describes an important change in people's lifestyles that has had an effect on the bears and their need to enter the communities (underlined ⓕ). Specifically, local residents used to go into the mountains to take care of the forests, but such practices have stopped in recent years. This statement shows that the local residents are afraid of bears, but some are aware that they are partly responsible for the bear encounters, as they contributed to the changes in the mountain environment. In other words, the residents have negative feelings toward the bears, but subconsciously, they accept the bears as animals that have long inhabited the neighboring mountains.

Changes in the community lifestyle are mentioned through these narratives about bears, as shown below. Mr. I, chief of district T in Tanto town, and Mr. J, a farmer, stated the following:

Mr. I: "Simply put, the quickest solution is to contact the local government. People used to work in the mountains to manage the forests. Today, people hardly go into the mountains to work. This is part of the issue. Our population is aging and decreasing. The mountains no longer sustain our lives. Our lifestyle has changed. Everyone works for a company. Agriculture does not pay."

Mr. J: "The aging of the population will accelerate during the next 10 years. People do not move to our community. This is good news for wildlife."

Mr. B, chief of district Y in Kami town, also mentioned the changes in the community:

"Our lifestyle has changed significantly. We no longer go into the mountains. We used to go into the mountains periodically to collect firewood. Today, we use gas and no longer have to obtain fuel from the mountains. The residents started to experience economic difficulties. They urged their children to study hard to help them move to urban areas and avoid the same difficulties. No one thought seriously about the survival of the community."

The Tajima region (northern part of Hyogo Prefecture), where these interviews were conducted, is one of the most rapidly depopulating and aging areas in the prefecture (Hyogo Prefecture 2005). Aging and depopulation is an especially serious issue in Tanto town; however, one district in the town has been making efforts to attract people from urban areas based on the catchphrase "nature, including bears." The statement below was made by Mr. K, who serves as a leader of district S and actively offers opportunities to experience agriculture to the members of the Nature Conservation Group who live in urban areas:

"Bears are valuable in that they represent a healthy forest ecosystem. Bears symbolize protection of the natural environment. I work with a Nature Conservation Group to protect the natural environment. What I have in mind is more than bears. Most members in that group live in urban areas. It is essential to raise their awareness. Paddy fields have been cultivated in district S to attract visitors from urban areas."

The community accepts people from urban areas several times a year to promote natural farming in collaboration with the local residents. In Hyogo Prefecture, district S is conveniently accessible from Osaka compared with other communities. Mr. K said, "I hope that young people in urban areas, even if only a few, will move to our community in the future."

4.3.2.2 Acceptance of Bears as a Part of Daily Life

The daily interaction between local residents and bears is discussed below. Six stakeholders, including a district chief and president of an agriculture association, were invited to hold a group discussion about bears for 2 h in district U in Tanto town.

Mr. D, a farmer: "Bears are very scary."

Mr. E, president of an agriculture association: "We are so scared of bears that we cannot go into the mountains."

Mr. F, a farmer: "Bears will come unless we kill them. We should kill bears immediately, as we kill mamushi (Japanese pit viper; *Gloydius blomhoffii*). The population of bears will increase if we release them."

In the beginning of the group discussion, Mr. D, Mr. E, and Mr. F often expressed negative feelings toward bears. At the end of the group discussion, they made the following statements.

Mr. D, a farmer: "In Tanto town, the residents are accustomed to encountering bears and wild boar. The residents take bear encounters for granted, so they do not panic when they encounter bears. They simply consider that it is the season."

Mr. G, a farmer: "Bears can eat tasty persimmon in season."

From these statements, it seemed that bear encounters have become part of the daily life of those local residents.

The following is a statement by Ms. H, a farmer in district U, whose persimmon trees in her garden at home are damaged by bears each year.

Ms. H, a farmer: "I can predict when bears will come by observing the persimmon trees. I am glad when the trees bear fruit, but I also can predict that bears will appear, so that make me sad. I am disappointed if the trees do not bear fruit, but it is acceptable because bears will not come. We talk a lot about such things in the neighborhood when we meet. In Tanto town, a bear entered the town hall building, and the staff evacuated to the second floor. I was told that a boar went into a coffee house in the town yesterday. People in our community joke that the bear came to the town hall to pick up a residence certificate, and the bear came to the coffee house to have a cup of coffee."

As Ms. H stated, local residents share information about bear encounters when they greet each other and make jokes about them. Bears are accepted as animals that are commonly observed in daily life. Ms. H suffers damage to her persimmon trees each year and thinks that bears should be "killed and reduced." In district Z of Kami town, bears and cubs used to be observed near residents' houses each year. A woman in the district said, "When the bear and cubs appeared, we observed them while having green tea." In Toyooka City, Hyogo Prefecture, people enjoyed watching the flight of the Oriental stork (*Ciconia boyciana*) until the 1930s; tea houses opened for bird watchers. People enjoyed watching Oriental storks, even though they were considered a nuisance and were, in many cases, chased from rice paddies because they were harmful (Kikuchi 2002). While bear encounters causing human injury are sometimes reported by the mass media, bears are part of the daily life of local residents, and sometimes they enjoy observing them.

4.3.2.3 Issues Arising from Frequent Bear Encounters

In a group discussion held in district U of Tanto town, when asked whether they report bear sightings to the town hall, some stakeholders responded, "I do not report it" or "I would have to call every day if I had to report it." Frequent bear encounters may prevent the residents from reporting sightings. Mr. L is a town hall official who is in charge of communities in the northern part of Kami town where there are relatively few bear encounters. He stated, "The residents rarely encounter bears. When they spot a bear, they report it to us." This statement suggests that people report bear encounters in communities where bear encounters are rare and that people are less likely to make a report in communities where bear encounters are frequent.

In a group discussion in district U, stakeholders discussed the management of persimmon trees, resulting in the following statements:

- "Persimmon trees cannot be wrapped with corrugated galvanized iron sheets if they have branches."

- "We put up barbed wire but it turned out to be ineffective."
- "We tried various measures, but bears come if they get hungry."

These statements represent the feeling of resignation; bears will come and cause damage, even if the residents take action.

Reluctance to make reports leads to delayed responses to bear encounters by the local government. Failure to wrap persimmon trees with corrugated galvanized iron sheets results in unexpected feeding by bears and leads to increased encounters. It seems that the appearance of bears is taken for granted, and increase in the number of conflicts (e.g., agricultural damage, mental stress) causes people to have even more negative attitudes toward this species.

Some communities ensure proper management of persimmon trees. One of them is district Y of Kami town. Mr. B, the district chief, stated the following:

> In district Y, officials visit each household to ask whether residents have cut down trees or wrapped them with corrugated galvanized iron sheets. In each community, all residents are expected to help each other. It is difficult for them not to go along with the others (i.e., not wrap persimmon trees with corrugated galvanized iron sheets).

The level of management for persimmon trees in a district may represent the strength of the ties between the residents and the district chief as well as that among the residents themselves in the community.

4.3.3 Discussion

In 2010, bears caused agricultural damage and human injury across Japan, and these incidents were reported by the mass media as a major issue. For some people, local residents in particular, bears are scary animals (Sakurai and Jacobson 2011). The number of encounters was particularly high in 2010. In Hyogo Prefecture, the number of bears captured by the end of the summer increased to above the average for other years. In Kami and Tanto towns, bears that were considered particularly dangerous were killed.

When local residents who have experienced bear encounters are asked about their perceptions, attitudes, and feelings toward bears, they first express negative feelings and hatred toward the bears, complaining about the local government that is responsible for the failure in management. A brief interview or questionnaire survey conducted by outsiders is likely to reveal only the negative perceptions of residents (Suzuki 2008). An interview of more than an hour revealed other thoughts about bears through the residents' narratives. We found that bears are considered (whether subconsciously or not) as neighbors in the mountains in the daily life of the community. Bear encounters may be very scary for local residents, but they seem to be accepted as an annual event in autumn.

The narratives about bears reflect the changing lifestyles in the communities, as perceived by the residents. That is, people no longer go into the mountains or manage the forests, resulting in a loss of vitality in communities and mountain villages. Meanwhile, one community has been making efforts to attract people from urban areas based on the theme of a natural setting inhabited by bears. The entire community is committed to revitalization because local residents understand the unique value of bears. The relationship between bears and residents has been built through various events, and this process is still underway (Maruyama 1999). Bears are considered as both enemies and neighbors in daily life. They also serve as indicators, reflecting the current status of the communities, and offering clues for community revitalization. The residents' narratives revealed that bears in the Tajima region are perceived as familiar animals; residents mostly have an ambivalent attitude.

Meanwhile, our survey found that people have become accustomed to bears in an undesirable manner, resulting in their reluctance to report bear encounters and, ultimately, a delayed response in the actions needed. In communities where people have become accustomed to bears, it is essential to raise awareness and work on damage prevention activities. In regard to the feelings of resignation that seem to accompany the persimmon tree management issues (i.e., "Bears would come even if we take action"), these concerns may have actually demotivated people from taking action. Within this context, damage prevention activities may be more effective if initiated and promoted by the district chief and local officials, as seen in the district Y where persimmon trees in the community have been managed properly.

Local residents view bears in various ways. It is essential to carefully analyze and understand these aspects when promoting community building efforts, so that productive activities can occur in a manner that enables residents to take pride in their communities. If citizens feel a supportive connection to their communities, they will also feel a desire to take voluntary action when needed, thereby creating a consensus toward solving community issues, like bear encounters.

4.4 Perceptions of Local Residents Regarding Black Bears: Quantitative Approach

4.4.1 Introduction

The previous section (Sect. 4.3) introduced the management of black bears in Hyogo Prefecture, as well as local residents' perceptions, as revealed by in-depth interviews. According to the Bear Conservation and Management Plan of Hyogo Prefecture (hereafter, the Management Plan), if a resident suffers damage by a bear, the prefecture gives permission to capture the bear. However, since they are considered an endangered species in the prefecture (as of 2012), captured bears are, in principle, released with aversive conditioning for behavior modification. Captured bears will be killed only when they have caused harm to people, the risk of human injury is high, or aversive conditioning is determined ineffective. As a result, between FY2003, when the Management Plan was developed, and FY2009, a total of 170 bears were captured for nuisance control, and a majority of them were released (only 20 were killed) (Hyogo Prefecture 2009). Compared with other prefectures, where the majority of captured bears are killed for nuisance control, the release rate in Hyogo is high and has led to an increase in the bear population in this prefecture (Yokoyama 2009). With regard to the conservation of bears, which is one of the purposes of the Management Plan, these aversive conditioning efforts have been somewhat successful, allowing for the use of non-kill methods.

However, it is not clear to what extent these efforts have resolved conflicts between bears and the local people. This is because, in order to resolve conflicts, the Management Plan focuses on actions to prevent encounters with bears (e.g., controlling bear-attracting objects), actions to prevent human injury, actions to prevent damage to agriculture and forestry, and actions to improve the reporting system for bear information; however, the majority of these actions are carried out solely by the initiative of local residents, and the state of progress is not evaluated. Results indicate that there were 1623 bear sightings/traces and 5 casualties in 2010, which were the worst figures recorded to date when compared with past records (Wildlife Management Research Center, Hyogo 2011, Ministry of the Environment 2011). Thus, it would seem that the conflicts have not yet been resolved.

In this study, a questionnaire survey of local residents was conducted to evaluate the progress of their conflict resolution approaches and to ascertain their current perceptions of bears. This survey asked to what extent the increased awareness and education activities performed by the government had affected local residents' behaviors toward preventing damage. The survey also aimed to identify the actions that local residents wanted the government to perform and then compared those with the actual actions taken. The findings revealed some hints regarding the direction of future plans to raise awareness and activities for future education, enabling managers to predict conflicts (differences of opinion) among stakeholders, as well as between stakeholders and managers (Vaske et al. 2010). The purpose of this study was to use the results to propose suggestions and policies that would encourage local residents to engage in damage prevention behaviors.

4.4.2 Methods

4.4.2.1 Questionnaire Items

Before conducting the questionnaire survey, stakeholders from the surveyed areas, including residents, hunters, and district chiefs, were interviewed (as shown in Sect. 4.3). These interviews found that concern over human-wildlife issues, including bear problems, was one of the major problems faced by communities, along with depopulation and aging of the remaining population. Survey questions were developed based on the findings from these interviews, and the issues were identified through discussions with prefectural/town office staff. First, a pilot survey was conducted using a small sampling of residents ($n = 5$) living within the study sites to identify any potential problem with the developed question. Consequently, I decided to include the following items in the survey:

1. The percentage of residents who implemented damage prevention interventions against bears.
2. Sources used by local residents for bear information and types of information that they would like to receive.
3. The percentage of damage prevention interventions implemented in each district against bears.
4. Actions that local residents want governments to implement when bears appear near human settlements.

With regard to (1), the survey questions identified how often residents reported bear sightings and/or traces of their presence and how well they managed bear-attracting objects, such as persimmon and chestnut trees. The Hyogo Prefectural government requested that residents report bear sightings and/or traces to their municipalities (Hyogo Prefecture 2009). Based on reports from local residents, the government typically investigates the causes of bear appearances and provides guidance on actions to prevent encounters with bears, including what to do if there are close encounters. Therefore, reports from local residents are important sources of initial information in cases of bear appearances. In addition, awareness-raising/education activities for managing bear attractants, such as persimmon and/or chestnut trees, are important activities that are carried out within communities (Hyogo Prefecture 2009). Although managers intuitively grasp the extent of reports from local residents and the degree of local residents implementing interventions to control bear-attracting objects, these reports and degrees of interventions have not been quantitatively studied thus far.

As for (2), municipalities (cities and towns) and their district chiefs publicize information about bear encounters to households through cable broadcasting, providing instructions for appropriate actions against bears. Municipalities also provide information about bears through public relations brochures and leaflets. Thus, by monitoring the local residents' dependence on these media sources, it is possible to determine which information would be most effective for future awareness efforts and education activities. In addition, clarifying the information that local residents want will help improve the quality of outreach campaigns conducted by administrative authorities.

Concerning (3), it is important to take action against wildlife, including bears, both individually and on a district-wide basis (Yokoyama et al. 2008). The survey questions were intended to help clarify differences in the levels of actions among respective communities and to identify regions that require focused guidance on awareness and education efforts regarding bears.

In reference to (4), the survey questions revealed which actions local residents want their governments to take when bears appear near human settlements. As of 2011, the Hyogo Prefectural government dealt with bear appearances based on four situation-specific categories (Hyogo Prefecture 2009):

Category 1: When a bear is seen in a specific area of concern (in the mountains, etc.) that is not a residential area. Action: Government collects information and alerts nearby residents of the bear.

Category 2: When people encounter bears in their residential areas (within human settlements, fruit farm, etc.). Action: Government removes bear-attracting objects and drives the bear out of the area.

Category 3: When a bear repeatedly appears and causes damage. Action: Government captures and releases the bear with aversive conditioning.

Category 4: When aversive conditioning is deemed ineffective or the risk of human injury is high. Action: Government kills the bear.

Grasping the difference between the actions that local residents want their governments to take and the actual actions taken will enable managers to assess local tolerance toward bears and to predict situations in which conflicts (differences of opinion) are likely to arise among local residents, as well as between stakeholders and managers (Vaske et al. 2010).

4.4.2.2 Study Sites

The questionnaire survey was conducted in Tanto town, Toyooka City, in the Tajima region of Hyogo Prefecture and in Muraoka-ku and Ojiro-ku, Kami town (Fig. 4.1).

Tanto town has a population of 5241 (Toyooka City 2010), and encounters with bears have been reported in many areas of the town. Kami town has a population of 21,010 (Kami Town 2010). Bears appear mainly in the southern mountain area of the town, and surveyed areas of Muraoka-ku and Ojiro-ku have the largest numbers of bear encounters in the town. In Kami town, five people had suffered casualties caused by bears in 10 years from 1999 to 2008 (Hyogo Prefecture 2009). The town office distributes leaflets to all households each year to alert people about the bears. In Tanto town, on the other hand, bear encounters occur frequently, but there are no records of human casualties in the past. The town does not distribute leaflets to all households to increase awareness and provide education on proper actions against bears as done in Kami town.

In June and July 2010, with the cooperation of the town offices and the district chiefs, a total of 2316 questionnaires were distributed to all 42 communities (1793 households) in Tanto town and to 16 communities (523 households) out of the total 74 communities of Muraoka-ku and Ojiri-ku in Kami town where there had been numerous bear encounters. The survey asked one person from each household, who was at least 18 years old and whose birthday was the closest to the day when they received the questionnaire sheet, to answer the questionnaire to avoid any imbalance resulting from age and gender.

4.4.2.3 Analysis

In analyzing the results of the questionnaire, unanswered questions were excluded, and valid responses were compiled. SPSS18 of IBM Japan, Ltd., was used for basic statistical processing.

On the district scatter diagram, which indicated actions taken against bears in each district, "experienced" (having seen/suffered damage) and "took action" (reported bears and wrapped persimmon and/or chestnut trees) were represented by scores of 0 (No) or 1 (Yes). The mean values of the respondents' "experienced" scores and "action" scores were calculated, respectively, and plotted on the diagram (e.g., the mean value of "took action" scores from a respondent who reported an encounter with a bear {1}, but did not take any action to wrap persimmon and/or chestnut trees {0} would be 0.5). We also conducted an analysis to confirm whether there was a correlation between having experience and taking action in each district ($p = 0.05$).

In addition, a pentagon graph was drawn to show the current state of each district by identifying specific characteristics based on five evaluation items:

① Degree of interventions implemented
② Experience of damage
③ Trust in government
④ Perception of bear-related risks
⑤ Problems faced by communities

Values from the district scatter diagram were used to evaluate numbers ① and ② on the pentagon graph. The mean values of responses to the survey questions were used to evaluate numbers ③ through ⑤ based on a five-scale score (1, Disagree; 2, Slightly disagree; 3, Neither agree nor disagree; 4, Slightly agree; 5, Agree) (Table 4.2).

Table 4.2 Questions used to draw pentagon graphs

Trust in government	Town government provided adequate information about bears
	Prefectural government provided adequate information about bears
	Town government provided opportunities to listen to people's concerns regarding bear issues
	Prefectural government provided opportunities to listen to people's concerns regarding bear issues
Perception of bear-related risks	Human-bear conflicts are increasing
	Concerned about the safety of children
	Concerned about agricultural damage
	Worried about walking outside in my neighborhood
Problems faced by communities	Reduction of manpower in the community due to aging, absence of young people, etc.
	Problems concerning public transportation means, hospitals, care facilities, etc.
	Problems concerning life in the community: An increase in abandoned farmland, funerals and festivals no longer held, etc.
	Pressure from outside the community: Damage by wildlife, disasters resulting from typhoons, etc.

4.4.3 Results

4.4.3.1 Sociodemographics, Experiences, and Degrees of Interventions Implemented

The response rate of the survey was 48.4% (n = 868) in Tanto town and 65.2% (n = 341) in Kami town. The percentages of male and female respondents were 75.0% and 25.0% in Tanto town and 63.7% and 36.3% in Kami town, respectively. According to census data, both in Tanto and Kami towns, the male-female ratio was approximately 50:50, and persons ≥65 years of age accounted for approximately 30% of the population (Kami Town 2009, Ministry of Internal Affairs and Communications 2005). Therefore, responses from the survey were given more often by males and elderly people in both towns, resulting in an imbalance in the respondents to the questionnaire (Table 4.3).

One question from the survey inquired, "How many times have you seen a wild bear in or around the community in the past ten years?" From both Tanto and Kami towns, more than 60% of the respondents answered that they had seen a bear at least once in the past 10 years, and more than 10% from both towns answered that they had seen a bear four or more times.

Another question from the survey inquired, "Do you report when you have seen a bear or traces of a bear?" To this question, 40% from Tanto town answered, "always" or "sometimes," which was significantly lower than the 59% from Kami town ($X^2 = 29.039$, $df = 2$, $p < 0.001$) (Table 4.4).

Table 4.3 Age of respondents

	Tanto town	Kami town
Age	Frequencies (%)	Frequencies (%)
Teenager	1 (0.1)	1 (0.3)
20s	13 (1.6)	2 (0.6)
30s	30 (3.7)	12 (3.8)
40s	92 (11.4)	20 (6.3)
50s	205 (25.1)	60 (19.0)
60s	221 (27.0)	91 (28.9)
Older than 70	255 (31.2)	129 (41.0)
Total	817 (100.0)	315 (100.0)

Based on Sakurai et al. (2012)

Table 4.4 Frequencies of respondents reporting bear sightings/traces

	Tanto town ($n = 625$)	Kami town ($n = 225$)	
	Frequencies (%)		X^2-test
Always	114 (18.2)	74 (32.9)	$X^2 = 29.039$, $df = 2$, $p < 0.001$
Sometimes	134 (21.4)	59 (26.2)	
Never	377 (60.3)	92 (40.9)	

Based on Sakurai et al. (2012)

Table 4.5 Reasons for not making a report

	Tanto town ($n = 494$)	Kami town ($n = 142$)	
	Frequencies (%)		X^2-test
It is not surprising to see a bear	200 (40.5)	41 (28.9)	$X^2 = 5.836$, $p = 0.016$
It is not mandatory to make a report	170 (34.4)	57 (40.1)	$X^2 = 1.337$, $p = 0.248$
Notification of a bear will scare tourists away and will have negative effects on tourism	26 (5.3)	6 (4.2)	$X^2 = 0.079$, $p = 0.779$
Others	143 (29.2)	48 (33.8)	$X^2 = 0.906$, $p = 0.341$

Based on Sakurai et al. (2012)

Among the reasons for not reporting, residents often responded, "It is not surprising to see a bear" or "It is not mandatory to make a report." These comments accounted for the highest percentage from Tanto (40.5%) and Kami towns (40.1%), respectively. In addition, the percentage of respondents answering "others" was 29.2% in Tanto town and 33.8% in Kami town. Many residents answered, "Nothing will happen, even if I make a report," "I saw the bear at night," or "I did not know to whom or to what organization I should submit the report" (Table 4.5).

In both towns, "district chief" or "town office" accounted for the highest percentage of responses to the question on whom respondents report bear sightings/traces to. In Kami town, the number of persons reporting to the district chief was significantly larger than that in Tanto town ($X^2 = 20.339$, df = 1, $p < 0.001$); in Tanto town, on the other hand, the number of persons reporting to the chairperson of agriculture or the police was significantly higher than that in Kami town ($X^2 = 13.219$, $df = 1$, $p < 0.001$; $X^2 = 7.020$, $df = 1$, $p = 0.008$, respectively).

More than 80% of the residents in both Tanto town (81.9%) and Kami town (80.8%) possessed persimmon and/or chestnut trees; the percentage of residents who suffered bear damage due to those trees was 40.3% in Tanto town and 40.0% in Kami town. Regarding the question on whether action was taken to prevent wildlife encounters due to the presence of the trees, 37.8% of the respondents in Tanto town answered that they took action, which was significantly less than the 66.4% who took action in Kami town ($X^2 = 61.493$, $df = 1$, $p < 0.001$). Among the reasons for not taking any particular action, most residents responded, "Persimmon and/or chestnut trees were damaged, but the damage was not very serious" (Tanto town = 46.9%, Kami town = 50.5%). Another slightly less common response from both towns was "I do not have enough manpower to take action" (Tanto town = 23.4%, Kami town = 24.5%).

4.4.3.2 Sources of Information for Local Residents Regarding Bears and the Information They Would Like to Receive

In Kami town, most residents (70.6%) obtained information about bear appearances from cable broadcasting, but in Tanto town, most residents (70.6%) received their information from other residents in the community (neighbors). Families, public relations brochures issued by municipalities, and newspapers also were mentioned by many respondents as a response to the choice "others."

Regarding information that local residents would like to receive, most residents in Tanto town answered, "Regions where bears appear, frequency, and details of damage," whereas most residents in Kami town answered, "Self-defense methods to prevent human injury." Following these responses, "Methods to protect crops and houses" was also mentioned by many respondents in both towns, and more than 20% of the respondents from Tanto town and more than 30% from Kami town selected "Locations of caught and released bears" as information that they wanted to receive (Table 4.6).

4.4.3.3 Degrees of Interventions Implemented to Prevent Bear Damage in Each District

Using the results described thus far, the characteristics of each district were clarified. These characteristics were plotted on a graph consisting of two measures, the degree of experience with bears (having seen bears and/or experienced damage) and

Table 4.6 Information local residents would like to receive

	Tanto town ($n = 814$)	Kami town ($n = 319$)	
	Frequencies (%)		X^2-test
Regions where bears appear, frequency, and details of damage	519 (63.8)	199 (62.4)	$X^2 = 0.133$, $p = 0.716$
Methods to protect crops and houses	349 (42.9)	145 (45.5)	$X^2 = 0.520$, $p = 0.471$
Self-defense methods to prevent human injury	481 (59.1)	202 (63.3)	$X^2 = 1.542$, $p = 0.214$
Government policies and budgets	108 (13.3)	64 (20.1)	$X^2 = 7.699$, $p = 0.006$
Ecology of bears	134 (16.5)	68 (21.3)	$X^2 = 0.960$, $p = 0.327$
Locations of caught and released bears	180 (22.1)	100 (31.3)	$X^2 = 10.015$, $p = 0.002$
Nothing particular	61 (7.5)	12 (3.8)	$X^2 = 4.695$, $p = 0.030$
Others	16 (2.0)	11 (3.4)	$X^2 = 1.575$, $p = 0.209$

Based on Sakurai et al. (2012)

the degree of action taken (having reported bears and/or wrapped persimmon and/or chestnut trees). Figures 4.4a and 4.4b show scatter plots of the different districts in Tanto and Kami towns, respectively. The value of the X-axis increases with the increase in the number of residents having some experience with bears, and the value of the Y-axis increases with the increase in the number of residents taking action. For example, although the value of the residents' experience with bears in district S (Fig. 4.4a) is not very high, their rate of action is high, and the district is, therefore, considered to be a model district. In district AF (Fig. 4.4a), on the other hand, many residents have had some experience with bears, and the possibility of suffering future damage is high; however, their rate of action is not very high. The results from the analysis of these data revealed no correlation between having experience and taking action in Tanto town ($r = 0.053$, $p = 0.744$), but a possible weak correlation in Kami town ($r = 0.456$, $p = 0.076$). For both towns, the correlation was not statistically significant.

In addition, Fig. 4.5 presents pentagon graphs showing the determined characteristics of district AF in Tanto town and district q in Kami town. Both of these districts presented the highest percentages of residents having some experience with bears within each respective town. There was no significant difference between districts AF and q in terms of experience of damage, residents' perceptions of bear-related risks, or the degree of problems that residents felt were facing their district; however, these figures show considerable differences between the two districts in terms of residents taking action, as well as their level of trust in the government.

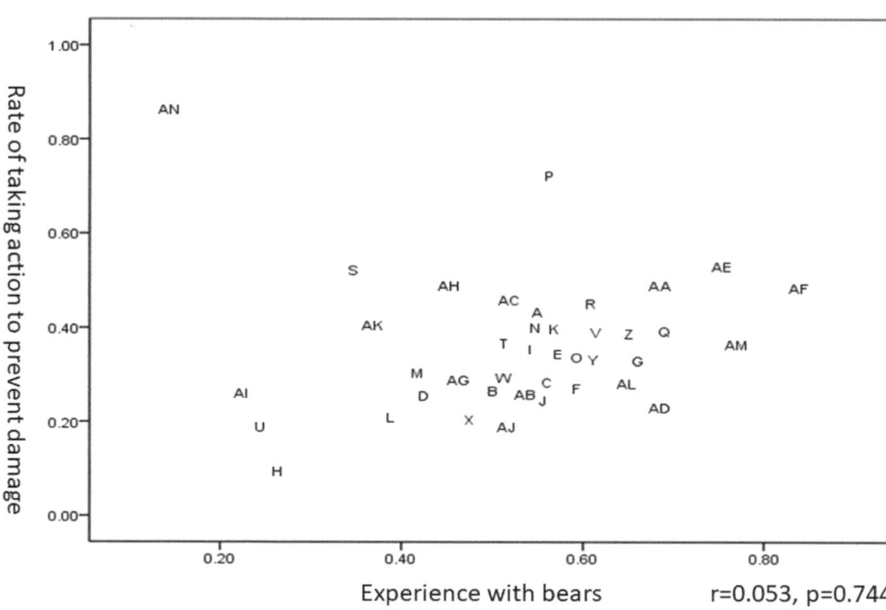

Fig. 4.4a Scatter diagram of districts in Tanto town. (Based on Sakurai et al. 2012)

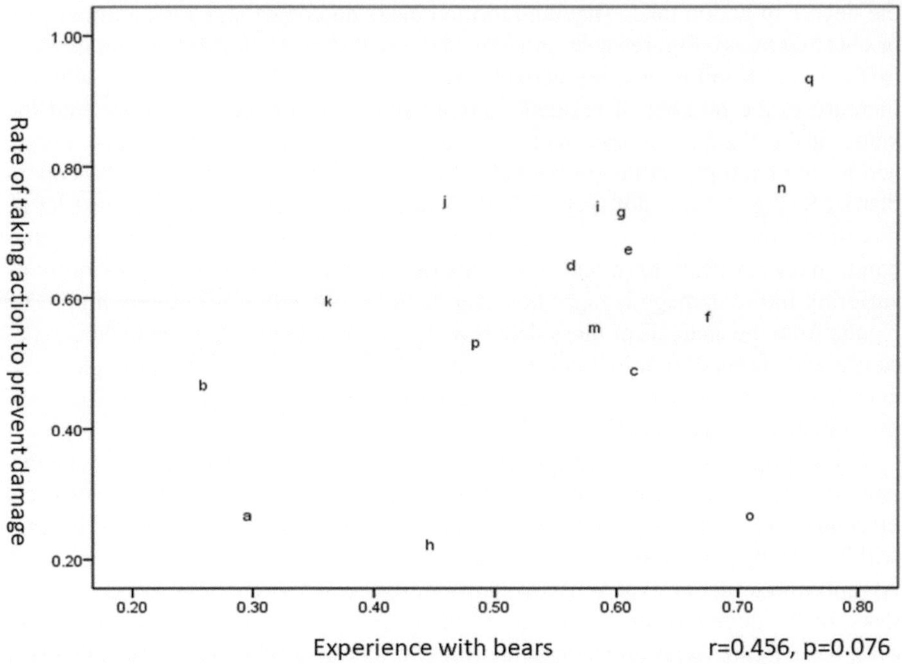

Fig. 4.4b Scatter diagram of districts in Kami town. (Based on Sakurai et al. 2012)

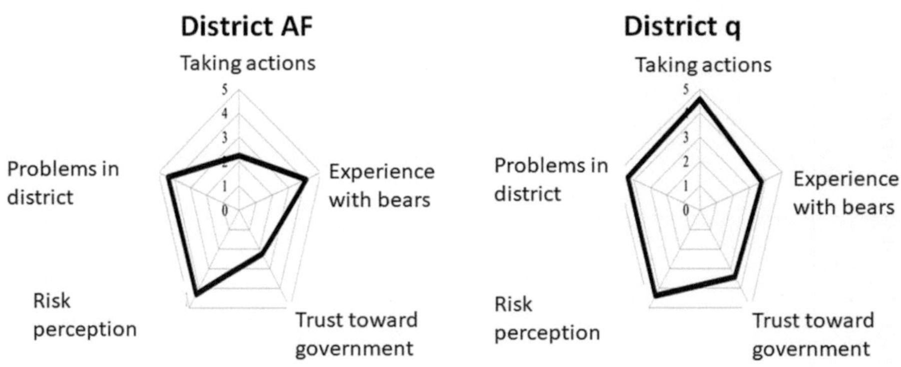

Fig. 4.5 Pentagon graph of the current state of districts AF (in Tanto town) and q (in Kami town). (Based on Sakurai et al. 2012)

4.4.3.4 Actions that Residents Want Local Governments to Implement When Bears Appear Around Human Settlements

Residents were asked what actions they wanted their governments to take in each of the following five situations:

1. When a bear is seen in the mountains
2. When a bear appears in the district (around human settlements)

3. When a bear appears on farmland and causes damage to crops or livestock
4. When a bear attacks a person in the mountains
5. When a bear appears in the district (around human settlements) and attacks a person

The results from these questions are shown in Table 4.7. Regarding (1) and (2), the attitudes of the residents were quite diverse, ranging from "do nothing" to "capture and kill"; however, their attitudes in numbers (3) to (5) did not vary widely; many residents wanted the bears to be captured and killed.

Table 4.7 Actions that local residents want their governments to implement

		Do nothing (%)	Threaten and drive the bear out (%)	Capture and release the bear without aversive conditioning (%)	Capture and release the bear with aversive conditioning (%)	Capture and kill the bear (%)
(1) When a bear is seen in the mountains	Tanto town ($n = 773$)	25.7	21.9	2.3	13.5	36.6
	Kami town ($n = 289$)	24.6	21.5	3.1	16.6	34.3
(2) When a bear appears in the district	Tanto town ($n = 764$)	4.8	23.6	2.4	18.2	51.0
	Kami town ($n = 286$)	3.5	16.8	4.5	23.8	51.4
(3) When a bear appears on farmland and causes damage to crops or livestock	Tanto town ($n = 770$)	1.8	7.9	1.8	25.8	62.6
	Kami town ($n = 289$)	1.0	6.9	2.4	27.0	62.6
(4) When a bear attacks a person in the mountains	Tanto town ($n = 762$)	2.6	4.2	1.0	6.8	85.3
	Kami town ($n = 289$)	3.5	3.8	1.7	6.6	84.4
(5) When a bear appears in the district and attacks a person	Tanto town ($n = 784$)	1.0	2.2	0.4	3.7	92.7
	Kami town ($n = 297$)	2.0	2.4	0.7	3.7	91.2

Based on Sakurai et al. (2012)

4.4.4 Discussion

4.4.4.1 Interpretation and Usage of the Results of the Questionnaire

When discussing the results of the questionnaire, we must consider the sociodemographics of the respondents and the fact that these respondents do not necessarily represent all residents throughout the entire region. In previous studies, perceptions of bears differed between males and females and between ages. For example, females generally did not want bears captured and destroyed (Agee and Miller 2009), while older residents held very negative attitudes about bears, feeling that they were dangerous animals (Canepa et al. 2008). Thus, caution is necessary when interpreting the results of surveys answered primarily by elderly males, as the increased negativity toward bears would not represent the perceptions of all local residents.

Although experience with bear damage was similar in both towns, both the rates of reporting and taking action were significantly lower in Tanto town than in Kami town, and there were significant differences in the reasons for not making reports. On the other hand, there were no significant differences in the reasons for not taking actions, like implementing interventions to prevent damage to their persimmon and/or chestnut trees. One of the most likely reasons for the observed difference in the rates of residents reporting and taking action between the two towns was the closeness of the relationships between the residents and the district chiefs. In Kami town, a larger proportion of residents claimed to report bear sightings to the district chief compared with residents in Tanto town, implying a closer, more trusting relationship between the residents in Kami town and their district chief. In addition, when bears appear, the district chiefs in both towns alert the residents through cable broadcasting and request that they take action in many cases (personal communication with a staff member at the Muraoka Regional Bureau, Kami town; personal communication with a staff member at the Tanto General Branch Office, Toyooka City); therefore, in a region where the district chief is more trusted, the residents would be more likely to comply with the district chief's instructions. The reporting will also increase the opportunities for the district chief and governments to provide guidance on actions against damage and consequently encourage the residents to take action.

Other reasons for the differences between the two towns in the rates of residents reporting and/or taking action for bear sightings could be the occurrence of human casualties and the distribution of leaflets by the town office to alert residents. In Kami town, there have been human casualties from bear encounters, and leaflets have been distributed by the town office, but in Tanto town (when this study was conducted), there had been no casualties recorded and no leaflets distributed by the town office.

In addition, a relatively large number of respondents cited lack of manpower and knowledge regarding the correct actions to take as reasons for not implementing interventions to prevent damage, such as wrapping persimmon and/or chestnut trees. The Tajima District Administration Office of Hyogo Prefecture, which has jurisdiction over the studied areas, has conducted several unique projects since 2009 to promote comprehensive actions to prevent wildlife damage, including the "Model Project to

Encourage Community-Based Actions Against Wildlife Damage" (Supporting Team for Wildlife Damage Prevention at the Tajima District Administrative Office of Hyogo Prefecture 2010). The "Wildlife Ranger Project for Damage Prevention" provides volunteer support in depopulated and aging regions that are lacking the manpower to implement interventions against nuisance wildlife (Sakurai 2019), and the "Community Wildlife Adviser Development Program" aims to train experts among the residential community to take action against nuisance wildlife. The issues described here may be resolved as these projects make greater progress.

Diverse results were obtained from the various districts regarding the progress of actions taken against bears. In both towns, there was no significant correlation between having experience with bears and taking action to prevent damage, and this study revealed that residents did not voluntarily take action, even if the damage increased. In Tanto town, where the rate of reporting was significantly lower than that in Kami town, the most frequent reason cited for not taking action was "It is not surprising that bears appear around human settlements." This shows that encounters with bears are daily occurrences and the residents are used to seeing them and feel there is little reason to take action and report the sighting [as explained in Sect. 4.3]. The most frequent reason cited for not taking action to prevent damage to persimmon and/or chestnut trees was "Persimmon and/or chestnut trees are damaged, but the damage is not very serious." This also implies an undesirable situation in which the residents have become somewhat indifferent to the bears, even though they suffer damage. The sense of resignation among the residents, who are used to encounters with bears and the resulting damage, along with a lack of manpower and lack of knowledge to take action, probably reflects their low motivation to respond to bear sightings, even though damage is increasing.

In addition, the pentagon graphs (Fig. 4.5), comparing the two districts with similar degrees of bear experience and damage, showed a considerable difference between the districts in terms of the residents' trust in the government. These results match previous studies revealing that trust in the government affects residential behaviors (Vaske et al. 2004). Accordingly, not only experience with bear damage but also other factors, such as being accustomed to bear appearances, could affect whether residents take action or not. This demonstrates the need for comprehensive policies and outreach actions against wildlife damage from the government, as mentioned above (e.g., Model Project) to motivate residents to engage in damage prevention behaviors.

The district scatter diagram, which plots the districts' experience with bear damage and their degrees of action, can serve as a resource for effectively selecting districts in which to carry out awareness-raising/education activities for residents in the future. In addition, we expect that providing local residents with a visual comparison between these communities will stimulate rivalry between the districts and synergistically promote action.

To conserve large mammals, it is necessary to improve local residents' tolerance toward the species and to obtain their consent for the necessary management approaches (Houston et al. 2010). Thus, the attitudes of local residents toward government actions are indispensable information for wildlife managers. In Hyogo

Prefecture, as of 2011, when a bear was seen in the mountains, administrative authorities would alert only nearby residents, initially not taking any particular action against the bear. This study revealed that local residents were divided in their desires for governmental actions under this situation; some believed the government should "do nothing," while others strongly believed in a "capture and kill" policy, indicating the difficulties in gaining a consensus among the population. When a bear appears in a community, Hyogo Prefecture aims to drive it out, but the prefecture's policy differs from the actions that local residents want. Thus, there is a high possibility of conflicts arising between managers and local residents.

4.4.4.2 Suggestions for Effective Policies and Applications to Actual Actions

Actions against nuisance wildlife, awareness-raising/education activities, etc. are conducted by governments, NGOs, and research groups throughout Japan and around the world where conflicts arise between wildlife and human beings (Conover 2002, Japan Bear Network 2007a, Waylen et al. 2010, Barlow et al. 2010). However, the impacts of specific actions and outreach programs in respective regions and communities have not been evaluated in depth (Gore 2004; Gore and Knuth 2006; Ferraro and Pattanayak 2006). Without advanced understanding of basic community information, prior to the initiation of awareness-raising/education activities or actions against damage, it is difficult to determine whether residents' perceptions and their degrees of action were initially high or were improved by implemented programs (Jacobson et al. 2006, Jacobson 2009). Thus, the results of this questionnaire survey provide useful information by establishing a baseline from which to measure and monitor the effectiveness of awareness-raising/education activities and damage prevention actions in the future. In addition, by quantifying the current status (residents' perceptions, degrees of action taken, etc.), it will be possible to set specific, realistic, and measurable goals for future awareness-raising/education activities and damage prevention actions (Jacobson 2009; Ernest et al. 2009; Carwardine et al. 2009). For example, by using this survey to determine that only 40% of residents were reporting bear sightings in Tanto town, Hyogo Prefecture was able to set the goal of increasing that rate to 80% over the next 2 years (personal communication with a staff member at the Agriculture, Forestry, and Fisheries Promotion Office of the Tajima District Administration Office, Hyogo Prefecture).

This quantitative questionnaire survey also yielded new findings that had not been discovered in other surveys. The chief of one district stated, "There are few bear appearances here"; however, when residents were interviewed, the questionnaire survey revealed that more than 70% of the local residents had seen bears in or around the community. The rate of reporting bear sightings in the community was less than 10%, indicating that interviews only with district chiefs or some stakeholders could not reveal the whole story (Hudenko et al. 2008); there was information available to help that community that even wildlife managers did not have.

Since this questionnaire survey was conducted, the Agriculture, Forestry, and Fisheries Promotion Office (Tajima District Administration Office, Hyogo Prefecture) has held seminars on taking actions against bears for local residents in 13 regions, including the surveyed regions. They stress the importance of reporting bear sightings to the government and sharing information with the community. As a result, government officials can take necessary actions according to the current determined status and needs of the region (personal communication with a staff at the Agriculture, Forestry, and Fisheries Promotion Office of the Tajima District Administration Office, Hyogo Prefecture). In the future, it will be important to monitor the effects of approaches and actions used, including the seminars implemented by the government in this region.

Throughout the survey, cooperation with nuisance wildlife management officials in the government was indispensable for choosing the tasks on which to focus, particularly for deciding which key information was most needed in the region before beginning the survey. Cooperation with wildlife managers is crucially important for researchers to determine and address research challenges in the field, as well as to promote more effective and efficient wildlife management policies.

It is also important for researchers to communicate the results of the survey to managers, local residents, and other community members, as well as for them to understand the results, enabling the local residents to take action by themselves (Enari and Suzuki 2010). To this end, clear and effective communication of the research results is important (Groffman et al. 2010), and researchers must master strong communication skills to make the type of research successful (Olson 2009; Jacobson 2009). In this survey, I worked to present the district scatter diagram, pentagon graphs, etc. in a visually concise manner, considering the subsequent reporting of the results to managers in the government, stakeholders, and local residents. Sharing these types of methods with other researchers will certainly initiate future successful endeavors; however, clearly communicating the results of surveys that reflect the current status of the region to managers and local residents, not to researchers, is the only way to establish a region of cooperative and involved residents and stakeholders that can coexist with wildlife (Jacobson et al. 2006; Groffman et al. 2010; Whitmer et al. 2010).

4.5 Implementation and Evaluation of a Community Bear Education Seminar: Observed Changes in Participants' Awareness and Behaviors After the Seminar

4.5.1 Introduction

As explained in Sects. 4.3 and 4.4, reducing conflicts between bears and local residents requires preventive actions within communities, such as appropriate management of persimmon and chestnut trees that attract wildlife and clearing bushes where bears may choose to hide. For these actions to be effective, consensus and commitment from local residents are essential. When local residents see bears, it is critical that they report these sightings to the government. However, as explained in Sect. 4.4, the results of the questionnaire survey of local residents in the Tajima region, where the largest number of bear encounters was reported in Hyogo Prefecture, revealed that the residents' rate of reporting bear sightings was far from satisfactory, even though these figures differed between districts: the rate of taking action to prevent damage was 37.8% in Tanto town, Toyooka City, and 66.4% in Kami town; the rate of people in the survey who responded that they reported bear encounters was only 18.2% in Tanto town and 32.9% in Kami town. The Bear Conservation and Management Plan includes the implementation of awareness-raising/education activities to develop a better understanding of bears and to encourage residents to engage in damage prevention actions (Hyogo Prefecture 2009).

In 2010, based on the results of the interview and survey conducted in the same year (Sects. 4.3 and 4.4), the Toyooka Agriculture, Forestry, and Fisheries Promotion Office (Tajima District Administration Bureau, Hyogo Prefecture) held seminars in 13 regions within its jurisdiction aimed at teaching residents how to take action against bears. These seminars were held as part of the awareness-raising/education activities outlined by the Management Plan to develop a deeper understanding of bears and encourage people to take action. Such extensive, in-depth seminars focusing on bears had not been held previously in the prefecture.

4.5.2 Purpose of the Study

While seminars and awareness-raising/education activities are conducted to instruct residents and tourists in various communities around the world about preventing damage by bears, there have been few studies on the effects of these efforts in terms of changing people's behaviors and reducing damage (Gore and Knuth 2006; Gore et al. 2006; Dunn et al. 2008, Baruch-Mordo et al. 2009). This is also true in Japan. The government, NGOs, etc. hold seminars and awareness-raising/education activities about bears throughout Japan (Japan Bear Network 2007b; Sakurai and Jacobson 2011); however, of the 16 prefectures that are promoting bear conservation and management plans, only Shimane, Hiroshima, and Yamaguchi Prefectures include evaluations of these activities in their plans (Ministry of the Environment 2010). Moreover, even these three prefectures are not conducting in-depth evaluations on the effectiveness of the programs (Personal Communication with staff, Shimane Prefecture Mountainous Region Research Center).

Therefore, a questionnaire survey aimed at seminar participants was carried out to evaluate the effects of the seminars on their perceptions and actions against bears. Using this survey, the effects and limitations of the seminars on bear awareness were clarified. In addition, a more effective approach for encouraging local residents to take action was discussed.

4.5.3 Method

4.5.3.1 Outline of Seminars

All seminars were held in the evening for representatives of communities and districts, school staff, police officers, and other interested local residents. Municipalities called for participation through handouts, cable broadcasting, etc. A total of 503 people participated in 13 seminars, each of which took about 2 h and explained the following items:

1. Details of appearances and capturing of bears
2. Requests to local residents to prevent encounters with bears and damage
3. Bear habits and behaviors that are important to know for avoiding damage
4. Details of bear habitats and government actions
5. Actions to prevent human casualties based on past cases
6. Actions to prevent bear appearances to maintain safety in communities

For item no. 1 of discussion, the staff of the Toyooka Agriculture, Forestry, and Fisheries Promotion Office explained the number of bear appearances in each region within the jurisdiction, details of the situations in which bears appeared (i.e., objects that the attracted bears), number of bears captured, and fates of the captured bears (killed or released). Regarding item no. 2, the staff explained the actions that must be taken by the local residents, such as reporting bear sightings/traces, sharing information with residents of the communities, and controlling bear attractants (trees and bushes). Staff of the Wildlife Management Research Center, Hyogo Prefecture, who are expert at dealing with bears, provided detailed explanations for item nos. 3–6. Concerning item no. 4, the staff explained that the bears inhabiting Hyogo Prefecture are designated as an endangered species and the prefecture, therefore, has formulated management policies centered around conservation. In addition, the staff explained that, based on these policies, the government has given permission to capture bears that cause damage and that captured individuals are to be released after aversive conditioning, in principle. Only bears whose behaviors could not be modified by aversive conditioning and those with a high risk for causing casualties would be destroyed. While discussing item no. 5 and reviewing past cases of human casualties, the wildlife management staff also explained what residents should do when they encounter a bear. The discussion of item no. 6 referred to the control of bear attractants and the importance of environmental maintenance and improvements in and around the communities. At the end of each seminar, time was allotted for a question-and-answer session.

4.5.3.2 Evaluation Method

In this study, the seminars were evaluated from three aspects:

- Quality (participants' comprehension of the contents of the seminars)

- Intention of the participants to take actions (whether the participants were motivated by the seminars)
- Implementation of actions by the participants (whether the participants actually took actions)

To evaluate the quality of the seminars and the intention of the participants to take actions, we conducted the first set of questionnaire surveys on-site at the end of each seminar. These were performed in all ten regions from August to October 2010. The following four items were asked in the first questionnaire:

1. Impression and comprehension of the seminar [whether they were able to acquire the information that they wanted, etc.]
2. Knowledge about bears [based on four items that residents should know to prevent damage from among the subjects in the seminar]
3. Actions taken before the seminar
4. Intention after the seminar to take actions [whether they had decided to report encounters with bears and engage in damage prevention behaviors as a result of participating in the seminar]

We evaluated the quality of the seminars based on the results of the responses to (1) and (2) and the intention of the participants to take actions based on the results of the responses to (3) and (4).

To evaluate the implementation of actions by the participants, we conducted a second questionnaire survey in late December 2010, after all seminars were complete. Since this survey was conducted by mail, the questionnaire was sent directly to those respondents of the first questionnaire who had given their names and addresses. The questionnaire was sent with a prepaid return envelope. The second questionnaire asked if, after the seminars, the residents had taken actions discussed in the seminars, and, if so, the reasons and details.

To evaluate the pure effects of the seminar, it was necessary to randomly extract a single group for which a seminar was conducted (treatment group) and a separate group for which a seminar was not conducted (control group) to compare the changes in perceptions and behaviors between the groups (Rossi et al. 2004). This random experiment model is considered to be the most effective and appropriate evaluation method, as it theoretically guarantees the two groups to be homogeneous and is highly objective. On the other hand, this method entails extra costs and time and is not always applicable to an actual societal situation, as it is accompanied by various constraints (Ryu and Sasaki 2010). In fact, subjects could not be extracted in a truly random manner for this study because the participants voluntarily joined the seminar, and thus, the seminar was evaluated with emphasis on changes in perceptions and behaviors of the participants.

4.5.3.3 Analysis

There could be regional characteristics, such as differences in the frequency of bear appearances, types of damage incurred by bears, and even differences in the ways that municipalities communicate with local residents. To take these characteristics into account, the results from the first questionnaires, administered after each seminar (venue), were analyzed. ANOVA was used to determine any significant differences between venues ($p \leq 0.05$). Significant results from the venue analysis are described below (Table 4.9), and overall aggregated results are described for those categories that showed no significant differences between venues (Table 4.9).

An open-ended question was posed to residents in the questionnaire survey regarding interventions implemented or those that residents were willing to implement; answers were categorized into seven groups:

- Managing garbage and food
- Maintenance of surrounding environments (i.e., cutting unnecessary trees, wrapping persimmon and chestnut trees, trimming shrubs, etc.)
- Sharing information about bear appearances and actions taken
- Actions to avoid encounters with bears (e.g., carrying a bell)
- Installing electric fences
- Installing trap cages
- Others

Regarding the knowledge about bears, we evaluated based on the percentage of correct answers to the four questions, as shown in Table 4.8.

4.5.4 Results and Discussion

Three hundred thirty-five people responded to the first post-seminar questionnaire. The rate of collection was 89%, and the average age of respondents was 60.3 years old (Table 4.9). A significant difference in the average age of respondents was found between venues ($F = 2.904$, $p < 0.01$). The second questionnaire was sent to 293 people, resulting in 227 responses (response rate: 77.5%).

4.5.4.1 Evaluation of the Quality of the Seminar

When evaluating impressions of the seminar, the percentage of respondents who answered "agree" or "slightly agree" was 94.6% for the item "I learned what I wanted to know about the habits of bears," 93.9% for the item "I learned how to minimize bear appearances," 91.2% for the item "I learned how to prevent human casualties by bears," and 76.9% for the item "I learned what the government is doing about bears." In addition, the average percentage of correct answers to the four questions concerning knowledge about bear habits was 93.2%. There were no significant differences between venues in terms of impressions about the seminar or the average percentages of correct answers to the questions concerning bear knowledge.

Because the majority of participants felt that they had acquired a better understanding of bears and what actions are required to help manage bears, it was deter-

Table 4.8 Questions and response forms from the first questionnaire

Principal category	Subcategory	Response form
Impression of the seminar	I learned what I wanted to know about the habits of bears	Five-scale scores (1, Disagree; 2, Slightly disagree; 3, Neither agree nor disagree; 4, Slightly agree; 5, Agree)
	I learned what the government is doing about bears	
	I learned how to prevent human casualties by bears	
	I learned how to minimize bear appearances	
Knowledge of bears	Bears have very good eyesight	Choose between two answers (1, Correct; 2, Incorrect)
	Bears are sensitive to sounds and smells	
	Bears like sweet food	
	Bears frequently move around during the daytime	
Behavior before the seminar	Before this seminar, when you saw a bear or trace of a bear, did you report it?	Choose between three answers (1, Always reported it; 2, Sometimes reported it; 3, Never reported it)
Intentions after the seminar	After participating in this seminar, have you decided to report bears sightings or their traces in the future?	Choose between three answers (1, Intend to always report it; 2, Intend to sometimes report it; 3, No intention to report it)
	After participating in this seminar, have you decided to take action on future sightings?	Choose between two answers (1. Yes; 2, No)
	Will you pass along to someone else the information that you have learned in today's seminar?	
	With what actions do you intend to follow through after the seminar?	Free response
	To whom will you communicate the details of this seminar?	

mined that the necessary knowledge was successfully communicated to the participants. The fact that there were no significant differences between the venues for these two items also indicated that the effects of the seminar were the same for each venue. Among the responses concerning impressions of the seminar, a better understanding of government actions was lower than that expressed about bear knowledge and actions needed to minimize bear damage. This could have been because the explanation of actions taken by administrative authorities was inadequate compared with the other items, and/or there could have been a difference between the actions that local residents expected from the government versus those that the government had actually taken (as shown in Sect. 4.4).

Table 4.9 Results of the first questionnaire with a specific focus on factors that were significantly different among venues

Venue	Average age [n]	Rate of reporting before seminar participation[a] (%) [n]	Rate of intention to take action after participation in the seminar[b] (%) [n]
Tanto town in Toyooka City	55.8 [26]	30.8 [26]	88.9 [27]
Takeno town in Toyooka City	61.8 [27]	52.6 [19]	73.1 [26]
Izushi town in Toyooka City	60.1 [47]	18.2 [44]	86.4 [44]
Kinosaki town in Toyooka City	59.6 [21]	56.0 [25]	68.2 [22]
Nakasuji-chiku in Toyooka City	58.5 [19]	70.0 [20]	77.8 [18]
Kamiyoshi-chiku in Toyooka City	55.8 [23]	36.4 [22]	72.7 [22]
Minato-chiku in Toyooka City	60.2 [25]	34.8 [23]	88.0 [25]
Other areas in Toyooka City	61.7 [34]	53.3 [30]	71.0 [31]
Muraoka-ku in Kami town	63.6 [40]	53.7 [41]	97.5 [40]
Ojiro-ku in Kami town	63.7 [18]	50.0 [24]	90.5 [21]

Based on Sakurai et al. (2013c)
[a]Rate of respondents who always reported seeing bears or their traces
[b]Rate of respondents who answered that they would take actions to prevent damage by bear after participating in the seminar

4.5.4.2 Evaluation of Participant Intentions to Take Future Action

Regarding the question on whether participants reported seeing bears or their traces, 86.2% answered that they would make a report after participating in the seminar; those who always had made a report prior to seminar participation were 43.8% (Table 4.10). There was a significant difference between venues in the rate of reporting before seminar participation ($F = 2.802$, $p < 0.01$) (Table 4.9). The highest rate of reporting was shown in Nakasuji-chiku, Toyooka City, where bear appearances were fewer than those in other regions, and in Ojiro-ku, Kami town where many appearances of bears were reported. As explained in Sect. 4.3, the rates of reporting tended to increase in regions with relatively few bear appearances, yet decreased in regions with many bear appearances because the local residents had become accustomed to bear sightings. However, these results revealed that the past frequencies of bear appearances did not always affect reporting rates. On the other hand, the intentions of the participants to report seeing bears or their traces greatly increased after participating in the seminar and did not show a significant difference between venues. As a result, the data show that it is possible to increase community motivation

Table 4.10 Comparison of rates for reporting sightings before the seminar and intentions to report after participating in the seminar

Question	Answer and ratio		
Did you report seeing bears or their traces before participating in the seminar? ($n = 274$)	Always (%)	Sometimes (%)	Never (%)
	43.8	23.4	32.8
Do you intend to make a report after participating in the seminar? ($n = 253$)	Always (%)	Sometimes (%)	Never (%)
	86.2	12.6	1.2

Based on Sakurai et al. (2013c)

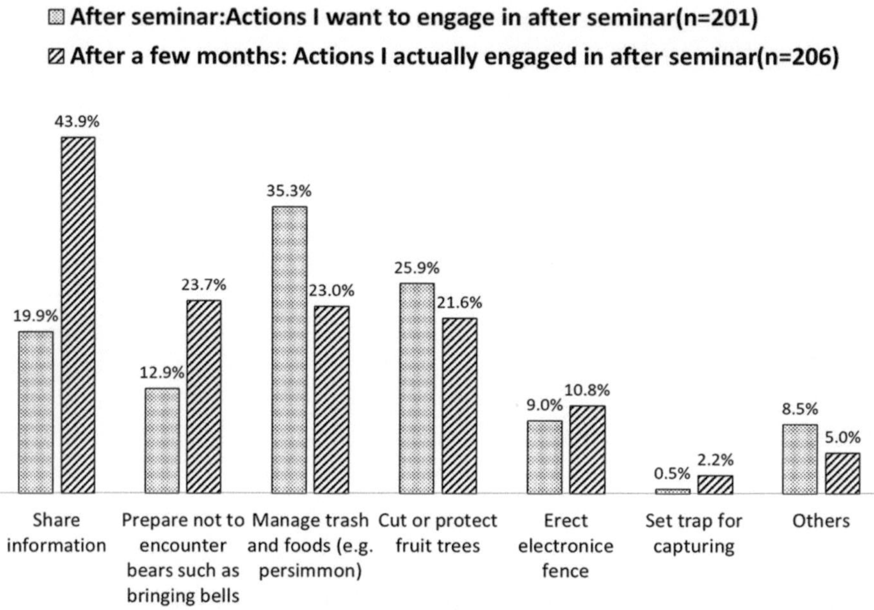

Fig. 4.6 Comparison of specific actions respondents intended to carry out after participating in the seminar (based on the first questionnaire) and actions actually taken after the seminar (based on the second questionnaire). (Based on Sakurai et al. 2013c)

levels, which were significantly different prior to the seminar, so that reports of sightings are comparable across the region.

Regarding whether the respondents intended to take actions after participating in the seminar, 82.6% answered “Participation in today’s seminar has encouraged me to carry out actions against bears from now on.” The highest percentages of specific actions taken included “manage garbage and foods” and “cut or protect fruit trees” at 35.3% and 25.9%, respectively (Fig. 4.6). A significant difference was found in intentions to take these actions between venues ($F = 2.099$, $p < 0.05$); it was high in Muraoka-ku and Ojiro-ku in Kami town where many appearances of bears are reported each year, but low in Kinosaki town in Toyooka City, and other regions with fewer bear appearances (Table 4.9). This is probably because the participants from regions with fewer bear appearances did not perceive any urgency for taking

Table 4.11 Based on the second questionnaire, degrees of actions taken to prevent bear damage after participating in the seminar

	Yes (%)	No (%)
Have you reported seeing bears or their traces since the seminar? (*n* = 101)	68.3	31.7
Did participation in the seminar encourage you to report the sightings? (n = 68)	66.2	33.8
Have you taken action to prevent bear damage since the seminar? (*n* = 219)	61.6	38.4
Did participation in the seminar encourage you to take action to prevent bear damage? (*n* = 138)	90.6	9.4

Based on Sakurai et al. (2013c)

these actions. In the future, it will be necessary to include an inspiring message in the seminars or other awareness-raising/education activities as a first step in regions with fewer bear appearances.

To the question "Do you think you will pass along what you learned in the seminar to someone else?" 95.1% of the respondents answered, "Yes," and there were no significant differences between venues. "Local residents" and "family members" accounted for the highest percentages (64.0% and 30.2%, respectively) of the people to whom the respondents would pass along what they had learned in the seminar. These results suggest that the information provided in the seminar was perceived as easy to share with neighboring residents and close family members.

4.5.4.3 Evaluation of Participants' Implementation of Actions

When responding to the question on whether participants had reported seeing bears or bear traces after the seminar, 68.3% of the respondents answered positively that they had made at least one report; the majority (66.2%) of these respondents stated that participation in the seminar encouraged them to report these sightings (Table 4.11). After participating in the seminar, 61.6% actually took actions, and most of those (90.6%) answered that participation in the seminar motivated them to do so.

Of the actions taken, "share information" accounted for the highest percentage (43.9%), followed by "prepare to alert bears of my presence (e.g., bells) (23.7%) (Fig. 4.6). There was a significant difference between participants' intentions to take certain actions after the seminar versus the actions that were actually taken. The percentages of people intending to "manage garbage and foods" and "cut and protect fruit trees" after the seminar compared with those who actually worked to achieve those goals decreased by 12.3% and 4.3%, respectively. It is possible that the respondents understood the need for these actions, but they were not easily achieved due to cost, lack of manpower, and/or difficulty in reaching a consensus within the community. On the other hand, sharing information and taking actions to avoid bear encounters (e.g., bells), those activities most frequently taken, required relatively little cost or manpower. These results imply that a one-time seminar can successfully help participants perceive a need for action, but it remains difficult to encourage participants to complete actions that require significant manpower.

Of the respondents, 94.5% answered that they did pass along to others what they had learned in the seminar. "Local residents," "family members," and "community officials (district chief, etc.)" accounted for the highest percentages (59.2%, 20.9%, and 18.4%, respectively) of the people to whom the respondents passed along their new knowledge. As explained in the Theory of Diffusion of Innovations, new ideas or norms first will be introduced by opinion leaders (those who strongly affect public opinion) or people in high positions within society (district chief, etc.) and then will gradually spread among other residents throughout the community (Rogers 1995). Many of the participants in the seminars were opinion leaders within their respective regions, such as community representatives, and most of them shared what they had learned in the seminar with family members and neighboring residents. Thus, it is expected that the important information will spread throughout the various regions, producing a ripple effect. To prevent damage by wildlife, it is important that not only specific individuals but also all residents of communities participate to take preventive and protective actions (Yokoyama et al. 2008, Sakurai et al. 2013a, b, c), so that a ripple effect can occur, strengthening the perceptions and behaviors of local residents.

4.5.5 Suggestions for Improving Programs

The actual rate of reporting significantly increased after attending the seminar, yet was still lower than the participants' intentions immediately after the seminar. An approach known as a "prompt" (a cue to exhibit correct behavior), which is frequently used in social marketing and other fields, may be an effective method to complement a one-time seminar like the one described within and could be used to increase bear reporting rates (McKenzie-Mohr and Smith 2006). For example, in an awareness-raising/education program for local residents about American black bears (*Ursus americanus*) in New York State (the USA), residents were given magnets describing the proper actions to avoid attracting black bears. Changes in residents' behaviors became apparent in the community because they were reminded by the prompt (magnets) as to which actions they should take, such as reducing the use of compost (Gore and Knuth 2006).

Regarding the implementation of actions, the current study revealed that taking actions requiring significant manpower was difficult for local residents. Field learning is an approach in which people become involved in the maintenance of the environment within the community, such as cutting bottom grasses, and it may be another effective addition to the seminars described within, thereby promoting an easier method for accomplishing more difficult actions (Jacobson et al. 2006). Personally experiencing the work required enables participants to accumulate knowledge and acquire confidence in carrying out the actions themselves. This method has been shown as an effective way to promote actions requiring manpower (Jacobson et al. 2006; Wals and Leij 2009; Yamabata 2011).

This study was the first step toward clarifying the role of seminars among various other awareness-raising/education activities for Asiatic black bears. Changes in people's perceptions and behaviors must not be temporary, and long-term monitoring of the continuity of these changes is necessary for continued success. It will also be important to investigate changes in people's perceptions and behaviors when other programs, including field learning and newspaper/handout education, are conducted in concurrence with seminars to develop more effective awareness-raising/education activities (Hayabuchi 2008).

In the field of wildlife management, only a small number of studies have been conducted to systematically evaluate the effects of awareness-raising/education activities, such as the ones described in this chapter; therefore, further studies in program evaluations are indispensable for successful wildlife management (Ferraro and Pattanayak 2006; Ueda et al. 2013).

References

Agee JD, Miller CA. Factors contributing toward acceptance of lethal control of black bears in Central Georgia, USA. Hum Dimens Wildl. 2009;14:198–205.

Barlow ACD, Greenwood CJ, Ahmad IU, Smith JLD. Use of an action-selection framework for human-carnivore conflict in the Bangladesh Sundarbans. Conserv Biol. 2010;24(5):1338–47.

Baruch-Mordo S, Breck SW, Wildson KR, Broderick J. A tool box half full: how social science can help solve human-wildlife conflicts. Hum Dimens Wildl. 2009;14:219–23.

Boykoff MT, Rajan T. Signals and noise. EMBO Rep. 2007;8(3):395–406.

Canepa S, Annis K, Kasworm W. Public opinion and knowledge survey of grizzly bears in the Cabinet Yaak Ecosystem: final report 2008. 2008. http://www.igbconline.org/Cabinet_Yaak_Survey_Final_101708.pdf. Accessed 10 Dec 2011.

Carwardine J, Klein CJ, Wilson KA, Pressey RL, Possingham HP. Hitting the target and missing the point: target-based conservation planning in context. Conserv Lett. 2009;2:3–10.

Conover MR. Resolving human-wildlife conflicts: the science of wildlife damage management. Boca Raton: Lewis Publishers; 2002. 418p.

Decker DJ, Brown TL, Siemer WF. Human dimensions of wildlife management in North America. Bethesda: The Wildlife Society; 2001.

Detjen J. The media's role in science education. Bioscience. 1995;45:S58–63.

Dunn WC, Elwell JH, Tunberg G. Safety education in bear country: are people getting the message? Ursus. 2008;19:43–52.

Enari H, Suzuki T. Risk of agricultural and property damage associated with the recovery of Japanese monkey populations. Landsc Urban Plan. 2010;97:83–91.

Ernest JA, Monroe MC, Simmons B. Evaluating your environmental education programs – a workbook for practitioners. Washington, DC: North American Association for Environmental Education (NAAEE); 2009. 171p.

Ferraro PJ, Pattanayak SK. Money for nothing? A call for empirical evaluation of biodiversity conservation investments. Public Library Sci Biol. 2006;4(4):482–8.

Garshelis DL, Steinmetz R (IUCN SSC BEAR SPECIALIST GROUP). *Ursus thibetanus*. In IUCN red list of threatened species v. 2012.2. 2008. http://www.iucnredlist.org. Accessed 21 Feb 2013.

Gore ML. Comparison of intervention programs designed to reduce human-bear conflict: a review of literature. Department of Natural Resources, Cornell University, HDRU Series No. 04-4. Ithaca: Cornell University; 2004.

Gore ML, Knuth BA. Attitude and behavior change associated with the New York NeighBEARhood watch program. Department of Natural Resources. Cornell University. HDRU Series No. 06-14. Cornell University, Ithaca, New York; 2006.

Gore ML, Siemer WF, Shanahan JE, Schuefele D, Decker D. Effects on risk perception of media coverage of a black bear-related human fatality. Wildl Soc Bull. 2005;33:507–16.

Gore ML, Knuth BA, Curtis PD, Shanahan JE. Education programs for reducing American black bear-human conflict: indicators of success? Ursus. 2006;17:75–80.

Groffman PM, Stylinski C, Nisbet MC, Duarte CM, Jordan R, Burgin A, Previtali MA, Coloso J. Restarting the conservation: challenges at the interface between ecology and society. Front Ecol Environ. 2010;8(6):284–91.

Hayabuchi Y. Diffusion effects of environmental education. Tokyo: Naknishiya Shuppan; 2008. (in Japanese).

Houston MJ, Bruskotter JT, Fan D. Attitudes toward wolves in the United States and Canada: a content analysis of the print news media, 1999–2008. Hum Dimens Wildl. 2010;15:389–403.

Hudenko WH, Decker D, Siemer W. Reliance on informants in wildlife management situation analysis – shortcut or shortcoming? Hum Dimens Wildl. 2008;13:459–70.

Huygens OC, Goto M, Izumiyama S, Hayashi H, Yoshida T. Asiatic black bear conservation in Nagano Prefecture, Central Japan: problems and solutions. Biosphere Conservation. 2001;3:97–106.

Hyogo Prefecture. Hyogo Prefecture red data book 2003. 2003. http://www.pref.hyogo.jp/JPN/apr/hyogoshizen/reddata2003/index.htm. Accessed 5 Dec 2017.

Hyogo Prefecture. 24: current situation of depopulated regions. 2005. http://www.sichouyouran.jp/document/list/24.html. Accessed 23 Sept 2010.

Hyogo Prefecture. Phase 2 black bear conservation and management plan. 2009. http://web.pref.hyogo.jp/contents/000140140.pdf. Accessed 2 March 2010.

Hyogo Prefecture. Addition to the exotic species list (black list) and red list. Hyogo: Hyogo Prefecture; 2011. (in Japanese). http://www.pref.hyogo.lg.jp/JPN/apr/topics/biodiversity/030_biotic_info/black_red_list_20110826.pdf. Accessed 10 Oct 2011.

Jacobson SK. Communication skills for conservation professionals. 2nd ed. Washington D.C: Island Press; 2009. 461p.

Jacobson SK, McDuff, MD, Monroe MC. Conservation education and outreach techniques. New York: Oxford University Press; 2006. 480p.

Jaeger CC, Ortwin R, Eugene AR, Webler T. Risk, uncertainty, and rational action. London: Earthscan Publications Ltd; 2001.

Japan Bear Network. Bears in Asia: current situation and future expectation. Ibaragi: Japan Bear Network; 2007a. (in Japanese).

Japan Bear Network. JBN symposium about bears and repot about workshop: summary of massive appearance of black bears in 2006 and suggestions from JBN. Ibaragi: Japan Bear Network; 2007b. p. 109.

Kami Town. Kami Town: beautiful mountain, rivers, and sea/town where people interact and coexist. 2009. http://www.town.mikata-kami.lg.jp/www/contents/1112674599328/files/kami-data2009.pdf. Accessed 15 Dec 2010. (in Japanese).

Kami Town. Kami Town. 2010. http://www.town.mikata-kami.lg.jp/www/toppage/0000000000000/APM03000.html. Accessed 16 Dec 2010. (in Japanese).

Kikuchi N. A study about the relationship between local people and the oriental white stork in the Tajima District, Hyogo Prefecture: reviewing the terms, "Turu" and "Konotori". J Environ Sociol. 2002;9:153–70. (in Japanese with English abstract).

Knight J. Waiting for wolves in Japan. An anthropological study of people-wildlife relations. New York: Oxford University Press; 2003.

Lombard M, Snyder-Duch J, Bracken CC. Content analysis in mass communication: assessment and reporting of intercoder reliability. Hum Commun Res. 2002;28:587–604.

Maruyama Y. Environmental history of human-wildlife conflicts. 1999. (in Japanese). http://staff.aist.go.jp/maruyama.yasushi/doc/lib.pdf. Accessed 3 Sept 2010.

McKenzie-Mohr D, Smith W. Fostering sustainable behavior: an introduction to community-based social marketing. Gabriola Island: New Society Publishers; 2006.

McQuail D. McQuail's mass communication theory. London: Sage; 2005.

Ministry of Internal Affairs and Communications. Census of 2005. 2005. http://www.e-stat.go.jp/SG1/estat/List.do?bid=000001005079&cycode=0. Accessed 21 Dec 2017. (in Japanese).

Ministry of the Environment. Materials about bears in Japan. Tokyo: Ministry of Environment; 2007a. (in Japanese). Available from http://www.env.go.jp/nature/yasei/kuma_manual/pdfs/mats.pdf. Accessed Oct 2007.

Ministry of the Environment. Methodological research of monitoring specified wildlife conservation and management plan: Fukushima Prefecture. Fukushima: Ministry of Environment; 2007b. (in Japanese).

Ministry of the Environment. A manual to respond to bear appearance: bears appearing from mountains. 2008. (in Japanese). http://www.env.go.jp/nature/yasei/kuma_manual. Accessed 12 Aug 2008.

Ministry of the Environment. Guideline to design the specified wildlife conservation and management plan (focused on bears). 2010. (in Japanese). http://www.env.go.jp/nature/choju/plan/plan3-2c/cov_index.pdf. Accessed 26 Dec 2017.

Ministry of the Environment. Human casualties caused by bears in 2010. Tokyo: Ministry of the Environment; 2011. http://www.env.go.jp/nature/choju/docs/docs4/injury-qe.pdf. Accessed 14 Apr 2011. (in Japanese).

Ministry of the Environment. Number of bears captured in 2016. 2016a. (in Japanese). https://www.env.go.jp/nature/choju/effort/effort12/capture-qe.pdf. Accessed 12 Jan 2017.

Ministry of the Environment. Number of casualties caused by bears in 2016. 2016b. (in Japanese). https://www.env.go.jp/nature/choju/effort/effort12/injury-qe.pdf. Accessed 12 Jan 2017.

Muter BA, Gore M, Riley SJ. From victim to perpetrator: evolution of risk frames related to human-cormorant conflict in the Great Lakes. Hum Dimens Wildl. 2009;14:366–79.

Olson R. Don't be such a scientist – talking substance in an age of style. Washington, DC: Island Press; 2009. 206p.

Onishi N. Chapter 3. Development of population and genetic structure: comparing east and West Japan. In: Tsubota T, Yamazaki K, editors. Bears in Japan: biology of brown bears and black bears. Tokyo: The University of Tokyo Press; 2008. (in Japanese).

Price V, Tewksbury D, Powers E. Switching trains of thought: the impact of news frames on readers' cognitive responses. Commun Res. 1997;24:481–506.

Riffe D, Lacy S, Fico FG. Analyzing media messages: using quantitative content analysis in research. Mahwah: Lawrence Erlbaum; 2005.

Rogers EM. Diffusion of innovations. New York: Free Press; 1995.

Rossi PH, Lipsey MW, Freeman HE. Evaluation: a systematic approach. 7th ed. Thousand Oaks: Sage; 2004.

Ryu Y, Sasaki R. Evaluation theories and techniques. Tokyo: Taga Shuppan; 2010. p. 217. (in Japanese).

Sakata H, Kishimoto Y, Seki K. In: Population estimate of Asiatic black bears, Wildlife Management Research Center, Hyogo, editor. Wildlife monograph of Hyogo, 3. Hyogo: Wildlife Management Research Center; 2011. p. 26–38. (in Japanese).

Sakurai R. Collaborative approach for coexistence with wildlife in rural regions of Japan. In: Frank B, Gilkman JA, Marchini S, editors. Human-wildlife interactions: turning conflict into coexistence. Cambridge: Cambridge University Press; 2019.

Sakurai R, Jacobson SK. Public perceptions of bears and management interventions in Japan. Human-Wildlife Interactions. 2011;5:123–34.

Sakurai R, Ueda G, Jacobson SK. Bears with various meanings recounted in the narratives of local residents: interviews at Tajima region of Hyogo Prefecture. J Rural Plan. 2011;30:399–404. (in Japanese with English abstract).

Sakurai R, Ueda G, Jacobson SK. Survey of residents regarding the Asiatic black bear at Tajima region of Hyogo Prefecture: designing and conducting questionnaire survey for policy application and management interventions. Wildl Conserv Jpn. 2012;12(2):33–46. (in Japanese with English abstract).

Sakurai R, Jacobson SK, Carlton JS. Media coverage of management of the black bear Ursus thibetanus in Japan. Oryx. 2013a;47(4):519–25.

Sakurai R, Matsuda N, Maruyama T, Jacobson SK. Overview of the model district program for reducing human-wildlife conflicts in Tochigi Prefecture. Wildl Hum Soc. 2013b;1(1):47–54. (in Japanese with English abstract).

Sakurai R, Ueda G, Jacobson SJ. Evaluation of the effectiveness of the community bear education seminar at Tajima region, Hyogo Prefecture: change of participants' awareness and behaviors after the seminar. Wildl Hum Soc. 2013c;1(1):29–37. (in Japanese with English abstract).

Siemer WF, Decker DJ, Shanahan J. Media frames for black bear management stories during issue emergence in New York. Hum Dimens Wildl. 2007;12:89–100.

Slovic P. Perceptions of risk. Science. 1987;236:280–5.

Supporting Team for Wildlife Damage Prevention at the Tajima District Administrative Office of Hyogo Prefecture. Report of 2009 model project for encouraging community-based actions against damage by wildlife. Hyogo: Tajima Administrative Office of Hyogo Prefecture; 2010. p. 28. (in Japanese).

Suzuki K. How can we resolve human-wildlife conflicts?: analyses on developing processes of wildlife problems focusing on local people's cognitive structure. J Environ Sociol. 2008;14:55–69. (in Japanese with English abstract).

Teel TL, Manfredo MJ. Understanding the diversity of public interests in wildlife conservation. Conserv Biol. 2010;24:128–39.

Toyooka City. Statistics of Toyooka City of 2009. 2010. http://www.city.toyooka.lg.jp/www/contents/1157942863609/index.html. Accessed 15 Dec 2010. (in Japanese).

Uchikoshi, A. Reports of monitoring research of 2007 specified wildlife conservation and management plan: Asiatic black bear. Nagano Prefecture, Nagano, Japan. 2007. (in Japanese).

Ueda G, Abe G, Sakata H. Evaluation of a program to improve the capture efficiency of feeding traps. Mamm Stud. 2013;53(1):31–42. (in Japanese with English abstract).

Vaske JJ, Timmons NR, Beaman J, Petchenik J. Chronic wasting disease in Wisconsin: hunter behaviors, perceived risk, and agency trust. Hum Dimens Wildl. 2004;9:193–209.

Vaske JJ, Beaman J, Barreto H, Manfredo MJ. An extension and further validation of the potential for conflict index. Leis Sci. 2010;32:240–54.

Wals AEJ, Leij TVD. Introduction. In: Wals AEJ, editor. Social learning: towards a sustainable world. Wageningen: Wageningen Academic Publishers; 2009.

Waylen KA, Fischer A, Mcgowan PJK, Thirgood SJ, Milner-Gulland EJ. Effect of local cultural context on the success of community-based conservation interventions. Conserv Biol. 2010;24(4):1119–29.

Whitmer A, Ogden L, Lawton J, Sturner P, Groffman PM, Schneider L, Hart D, Halpern B, Schlesinger W, Raciti S, Bettez N, Ortega S, Rustad L, Pickett STA, Killilea M. The engaged university: providing a platform for research that transforms society. Front Ecol Environ. 2010;8(6):314–21.

Wildlife Management Research Center, Hyogo. 2011. Sightings of Asiatic black bears. Hyogo, Japan. http://www.wmi-hyogo.jp/bear_info/bearinfo_h2309.pdf. Accessed 28 Nov 2011. (in Japanese).

World Association of Newspapers. World's 100 largest newspapers. 2005. http://www.wan-press.org/article2825.html. Accessed 20 Mar 2011.

Yamabata N. Effect of chasing away by village on the home range and appearances of a macaques group: verification in 7 area of Mie Prefecture. J Rural Plan. 2011;30:381–6. (in Japanese with English abstract).

Yokoyama M. 6. Asiatic black bear: reviving from a danger of extinction. In: Kawai M, Hayashi Y, editors. Doubutsutachino Hanran (rebellion of wildlife). Tokyo: PHP Research Inc.; 2009. p. 129–58. (in Japanese).

Yokoyama M, Sakata H, Morimitsu Y, Fujiki D, Muroyama Y. Current status of and perspectives on the specified wildlife conservation and management plans and monitoring for Japanese black bears in Hyogo Prefecture, Japan. Mamm Sci. 2008;48:65–71. (in Japanese with English abstract).

Chapter 5
Human Dimensions Studies on Programs for Reducing Human-Wildlife Conflicts in Tochigi Prefecture, Japan

Abstract This chapter introduces a series of human dimensions studies conducted at Tochigi Prefecture, with a specific focus on the Model District Program. In Sect. 5.1, I explain the background of the Model District Program. The results of the interviews and surveys conducted among local residents to understand the impacts of this program are explained. The two issues faced by the program are then highlighted: the sustainability of the intervention and the broad participation of residents. Section 5.2 explains the results of the surveys conducted among local residents of two model districts. Based on social-psychological theories developed in the USA (the Theory of Planned Behavior and the wildlife acceptance capacity model), the study revealed that social norms significantly explained residents' behavioral intentions, while risk perceptions significantly explained residents' capacity for wildlife acceptance. Section 5.3 examines some comparative studies carried out over different years and districts. The results revealed in model districts where participatory interventions were implemented, residents were more likely to engage in damage prevention measures. They were also more satisfied with government performance than in neighboring districts not subject to interventions. Over time, however, the number of the participants in model district interventions decreased, indicating the importance of maintaining motivation among residents.

5.1 Overview of the Model District Program for Reducing Human-Wildlife Conflicts

5.1.1 Introduction

Located in Central Japan, 70% of the land area in Tochigi Prefecture is either forest or farmland (Tochigi Prefecture 2010). The populations of certain wildlife species such as deer and wild boars are increasing due to factors including depopulation, an

Please bear in mind that this chapter was written based on academic studies, and the resulting papers written by the author. Section 5.1 is based on Sakurai et al. 2013a; Sect. 5.2 is based on Sakurai et al. 2014a; and Sect. 5.3 is based on Sakurai et al. 2014b.

R. Sakurai, *Human Dimensions of Wildlife Management in Japan*, Ecological Research Monographs, https://doi.org/10.1007/978-981-13-6332-0_5

aging society, and the abandonment of agricultural land resulting in the expansion of wildlife habitats. A decline in the number of hunters has also brought about an increase in wildlife numbers. Thus, agricultural damage by wildlife in the prefecture is a serious problem. In 2015, there were 2,975 tons of annual agricultural damage, costing 371.99 million yen (about 3.7 million USD) (Ministry of Agriculture, Forestry and Fisheries 2015). In terms of the animals responsible, wild boars caused about 52% of this damage, deer about 9%, and macaques about 7% (Ministry of Agriculture, Forestry and Fisheries 2015).

In Tochigi Prefecture, measures to prevent agricultural damage by wildlife are taken mainly by the Rural Development Division of Department of Agricultural Administration, while wildlife management is handled primarily by the Nature Preservation Division of the Department of Environment and Forestry. At the city or town level, departments in each city government or town office are responsible for damage countermeasures and wildlife management.

In the 1970s, much of the prefecture's independent wildlife research related to hunting resources in the form of wildfowl, including mountain pheasants (*Syrmaticus soemmerringii*) and common pheasants (*Phasianus versicolor*). Beginning in the 1990s, the appearance of deer, wild boars, black bears, macaques, and other large mammals became more common, with the associated agricultural damage (Niibe et al. 2008). Thus, research into these species was commenced. In 2006, a Specified Wildlife Management Plan (hereafter "Specified Plan") was implemented for these four large mammals (deer, wild boars, black bears, and macaques). The plan was independently formulated and enacted by the prefecture, which has since taken steps to control animal populations appropriately and to mitigate human-wildlife conflicts (Niibe et al. 2008).

In FY2008, the prefecture introduced the Prefectural Tax for Healthy Forest Creation, the goal of which was to allow residents to help nurture the prefecture's forestry. Projects financed by this tax include the Satoyama Woodland Improvement Project aimed at reducing human-wildlife conflicts. For this project, the prefecture pays funds to regional residents (or to local residents' associations, etc.) via city and town governments. These funds cover the cost of improvement work such as clearing away unnecessary trees or cutting back brush adjoining paddies, upland fields, or residential neighborhoods. Financing extends also to maintenance work for a 4-year period after these improvements have been carried out.

Human-wildlife conflicts cannot be solved if only a limited group of highly aware people take steps to deal with the issues at hand. If neighbors neglect to manage raw garbage and other materials that attract wildlife, for example, conflicts will continue to occur. Hence, it is important to increase the level of awareness among residents and to ensure that they participate in countermeasures (Yamabata 2011). In Tochigi Prefecture, an overall framework for wildlife management (the Specified Plan) has been established, and concrete damage countermeasures and awareness-raising activities in the regions where damage occurs have generally been conducted by city and town authorities (Maruyama 2005). However, details about region-specific interventions, including success rates, are yet to be consolidated.

To mitigate wildlife conflicts, the Tochigi Prefecture Environment and Forestry Department realized it is vital to create an environment that does not attract wildlife. Similarly, it is necessary to implement community-based interventions by gaining the understanding and cooperation of local residents. To this end, the Model District Program for Reducing Human-Wildlife Conflicts (hereafter "Model Program") was launched by the prefecture in FY2010. This project was intended to create a regional implementation system allowing local residents to independently and sustainably undertake damage prevention activities. Its twofold aim was, first, for local residents, aided by experts, to clarify the state of damage to their own locality and to take appropriate action to reduce it and, second, to increase their awareness of damage prevention countermeasures. The program was also expected to have a ripple effect. It was anticipated that the activities of model district residents would motivate those in other districts to take similar countermeasures. The following are typical steps of one such Model Program.

1. A city or town recommends a district that meets the requisite conditions (that frequent human-wildlife conflicts occur and that the local residents are unified in seeking countermeasures) to the prefecture.
2. The prefecture holds a meeting where it explains the Model Program to concerned parties in the recommended district. If the district agrees to the program conditions, it is designated as a model district.
3. Model Program activities commence. These include, about once every 6 months, lecture meetings held by experts, field trips to check wildlife damage sites with the participation of residents, and cutting back of underbrush and unnecessary fruit trees around residential land and farmland.

Previous studies have examined the ways that communities adopt and implement activities involving the entire region. These included the analyses of the factors behind the continuance of regional environmental conservation activities (Kimura et al. 2008) and the factors encouraging nonfarmers to participate in the maintenance of agricultural-use irrigation channels (Yamamoto and Nagasawa 2010; Honda 2011). Studies regarding efforts by a region to prevent human-wildlife conflicts include those by Kinoshita et al. (2009) and Yamabata (2009). However, there has been no research that utilized both qualitative (interviews, etc.) and quantitative (questionnaires, etc.) approach to reveal the factors behind the continuance of the project and no evaluation of residents' reactions to these activities.

It is difficult in terms of both time and money for governments to conduct project evaluations, as their energies and resources are devoted primarily to implementing policies and actions (Ferraro and Pattanayak 2006). And the situation in Tochigi Prefecture is no different. While awareness-raising/education activities or damage countermeasure projects have been undertaken sporadically by administrative bodies in specified regions, a detailed evaluation of these projects examining changes in residents' awareness or their willingness to intervene has not been carried out (personal communication with a staff member at Nature Preservation Division of Tochigi Prefecture).

Hence, this study was undertaken to clarify the conditions and potential problems relating to the Model Program established in Tochigi Prefecture since 2010. The aim was to examine residents' reactions and impressions and to propose measures to increase effectiveness of the Model Program.

5.1.2 Study Sites and Methods

The study focused on two districts participating in the Tochigi Prefecture Model Program: Momurahonden in Nasushiobara City and Myojin in Nikko City (Fig. 5.1, Table 5.1). Interviews were conducted with the heads of residents' associations, farmers, and other stakeholders. Questionnaire surveys were distributed to those taking part in a lecture meeting or an on-site training session held in conjunction with the program. One-month follow-up surveys (interviews) with the questionnaire respondents were conducted to determine whether changes in behavior had occurred.

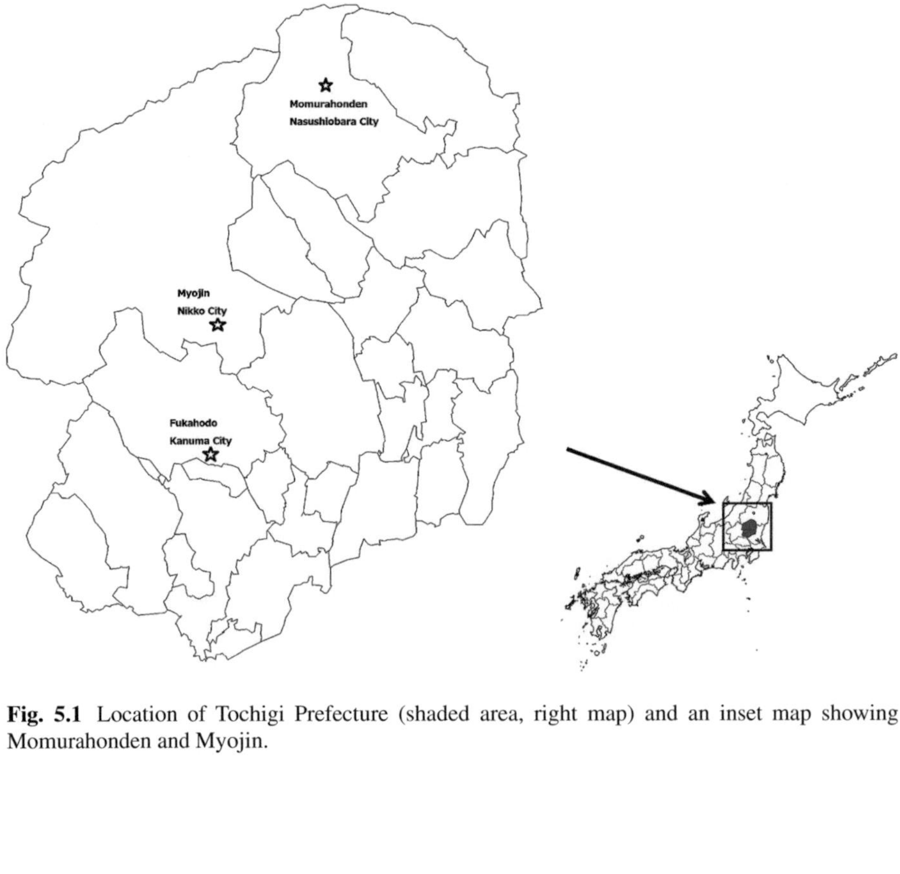

Fig. 5.1 Location of Tochigi Prefecture (shaded area, right map) and an inset map showing Momurahonden and Myojin.

Table 5.1 Characteristics of the three districts where the Model Program was implemented

	Momurahonden, Nasushiobara City	Myojin, Nikko City	Fukahodo, Kanuma City
Number of households	57	538	189
Average age	42[a]	50	56[b]
Typical wildlife species that cause damage	Macaques, black bears, wild boars	Wild boars, macaques, deer	Wild boars, macaques, deer
Year when the Model Program started	2010	2011	2010
Organizations that deal mainly with wildlife issues	Residents' association/ Forest Protection Society	Wild Boar Countermeasure Committee	FukahodoSatoyama Protection Association
Achievements of the Model Program[c]	Good opportunities to carry out community-based interventions/ damage caused by macaques decreased	Good opportunities to share ways to improve the effectiveness of countermeasures (electric fences, etc.)	Good opportunities to increase nonfarmers' awareness of wildlife issues
Challenges of the Model Program[c]	Sustaining activities/ lack of associations or people prepared to take the initiative in carrying out activities	Fostering interventions in the eastern part of district where many new residents live	Implementing interventions led by local residents

This table was created based on Sakurai et al. (2013a)
[a]Based on Nasushiobara City (2011)
[b]Based on a survey conducted by the author in 2011
[c]Based on interviews with residents

5.1.3 Results and Discussion

5.1.3.1 The Case of Momurahonden in Nasushiobara City

Since the 1980s, a troop of macaques has been appearing around the district of Momurahonden, in northeastern Nasushiobara City. Black bears and deer have also appeared and caused agricultural damage. Recently, intrusions by wild boars have similarly been confirmed.

Model District operations began in February 2010, when a meeting was held by the local residents' association, officials from administrative bodies (city and prefecture), and researchers from Utsunomiya University. In March 2010, field trips were carried out, mainly by local residents, to confirm the state of damage in the area, to identify wildlife intrusion routes, and to prepare a village inspection map. Later, a lecture meeting was held to consider ways to drive out the macaques. Unnecessary fruit trees that attract wildlife were subsequently cut down, with the participation of local residents (Fig. 5.2).

Fig. 5.2 A lecture meeting about interventions to prevent damage by macaques (left). The cutting down of unnecessary fruit trees that attract wildlife (right). (All the photos were taken by the author at Momurahonden)

In July 2011, a lecture meeting about wild boars was held, and a questionnaire survey was conducted to determine the residents' impressions of the meeting. The average age of the 17 respondents (response rate, 89.5%) was 58.2 years. Seven of these had not been engaged previously with activities to prevent damage by wildlife. Regarding the contents of the lecture, approximately 90% of respondents indicated that they had learned about the ecology of wild boars, and 90% stated that they had learned ways to prevent damage by wild boars. All the respondents answered that they had decided to take damage countermeasures. These included removing underbrush and cutting grass (n = 5) or undertaking interventions as a region (n = 2).

Follow-up interviews (n = 5) were conducted about 1 month after the lecture. These revealed many positive opinions about the Model Program, such as "The appearance of macaques has fallen (since the start of the prefecture's Model Program)" and "Clearing and tidying underbrush has been effective." However, some of the respondents had negative views about the program's future prospects, such as "We did it because the prefecture sent people out, but if they stop coming, it will not continue. That is because there will be nobody to lead us."

In Momurahonden, the head of the residents' association changes annually, making it difficult to carry out continuous activities. For this reason, a resident commented, "Wouldn't the program be more sustainable if not only the residents' association but also the Forest Protection Society took part?" The Forest Protection Society is the organization that manages the forests in the Momurahonden district. It has around 40 members, including proprietors of jointly owned forests, operators of lumber companies and forest-related businesses, and other residents of the district.

In Momurahonden, the lecture meeting and damage interventions led by the residents increased awareness among participants and encouraged them to act. Many residents stated that, "It was a good opportunity," and a number reported that the project had reduced the frequency of intrusions by macaques. However, in the future, it will be necessary to address the following key concern: "To what degree can the region continue with this intervention?"

Fig. 5.3 The electric fences (left). Sign warning not to touch the electric fences posted by the Wild Boar Countermeasure Committee (right) in Myojin. (All the photos were taken by the author)

5.1.3.2 The Case of Myojin in Nikko City

Myojin is located in southeastern Nikko City. In FY2008, 3 years before it was designated as a model district, electric fences were installed along the foothills in the western part of the district. Fifty percent of the costs of this installation were subsidized by the Department of Agricultural Administration of Tochigi Prefecture (Fig. 5.3). At this time, Mr. A, who served as the head of the district's residents' association, took the opportunity to launch the Wild Boar Countermeasure Committee. Then, in FY2011, the Prefectural Nature Preservation Division designated Myojin as a Model District. In June of the same year, administrative officials (city and prefecture), the head of the district residents' association, and the chairman of the Wild Boar Countermeasure Committee conducted preliminary discussions. In August, they invited researchers from Utsunomiya University for discussions. A wild boar countermeasure lecture meeting and an on-site training session were also held. At the lecture meeting, information was put forward about where the electric fence could be effectively built. Methods of managing the fence based on earlier damage interventions in Myojin were also proposed, followed by a question and answer session. After the lecture meeting, the participants visited sites where electric fences had been installed previously. The results of the questionnaire survey immediately following the lecture meeting were as follows. The average age of the 19 respondents (response rate, 100%) was 65.5. Of these, 90% had taken part in wildlife damage prevention activities previously. All the respondents answered they had learned about the ecology of wild boars and how to prevent damage. More than 70% stated they had decided to undertake damage prevention activities following the lecture meeting, including installing electric fences and cutting the grass around existing fences.

Approximately 1 month later, a follow-up interview of named respondents (n = 6) revealed that many considered the lecture meeting to have been somewhat effective. They stated also damage prevention activities by local residents had already been undertaken in Myojin, even before the lecture meeting.

Myojin is a district that experiences human-wildlife conflicts. In 2008, households in the foothill areas on the western side of the district (approximately 40 people) worked together for several days to install 1.8 km of electric fences. The city government supplied the building material and the local people prepared and installed the fences. Following this, residents of the western part of the district, led by the chairman of the Wild Boar Countermeasure Committee, undertook to cut back the underbrush every year. They took turns also using a voltmeter to carry out monthly checks of the fence current and, when necessary, to remove grass from around the fence.

These monthly inspections continue to this day (November, 2011). While the Wild Boar Countermeasure Committee led these activities, there was also a feeling of solidarity within the district. Farmers and nonfarmers alike came together to install the fences. This camaraderie was highlighted by an attendee of the lecture meeting, Mr. B, who said, “In Myojin, we get along very well. Nobody said that installing the wires was ‘not my concern’. Everyone took part, and will assist in sustaining these wildlife interventions.” In the first year of the Satoyama Woodland Improvement Project that started in FY2009, a company was contracted to cut back the underbrush. As it was considered preferable that the city government pay the maximum amount to the beneficiaries, however, the local residents played a leading role in underbrush removal from 2010 onwards. In Myojin, residents notify the Wild Boar Countermeasure Committee whenever damage occurs, as they are aware of the committee’s role in organizing local damage prevention interventions.

In the western part of the district, electric fences have effectively reduced the degree of damage. A new challenge appeared in 2011, however, when wild boars started to emerge from the forest in an area along the tracks of the Nikko Line of the Tobu Railway in the east of the district. In contrast to the west side of Myojin, which is home to many long-term residents, many people living in the east side of Myojin are recently arrived office workers. Hence, they are not as aware of the wildlife problems and not as motivated to engage in damage prevention interventions. Therefore, a future challenge is to find ways of enhancing the awareness of and the willingness to intervene by these new eastside residents.

5.1.4 Two Issues Facing the Model District Program

5.1.4.1 Sustainability of the Intervention

As shown by this study, one issue facing this project is the sustainability of the intervention. In Myojin, the Wild Boar Countermeasure Committee, represented by the chairman of the residents’ association at the time, Mr. A., was launched when the electric fences were installed. This suggests that new organizations attached to residents’ associations are required to manage and operate funds whenever a Satoyama Woodland Improvement Project is introduced in a district.

Conversely, when Momurahonden was designated as a Satoyama Woodland Improvement Project in FY2010, organizations specializing in human-wildlife conflict interventions were *not* established under the residents' association. Instead, the Forest Protection Society was contracted to implement the project. It is possible that the presence of an organization responsible for managing forests in the district diminished the residents' sense that they needed to form a new organization. However, the Forest Protection Society is not an organization that specializes in human-wildlife conflict interventions. Thus, the question arises: "How will the intervention be continued once the Model Program has ended, eliminating the active role played by the prefecture?" Residents have conflicting views about whether it will be possible for the Forest Protection Society to continue human-wildlife conflict interventions or whether it will be necessary to form a separate organization under the umbrella of the residents' association. To date, a definitive answer remains elusive.

Kimura et al. (2008) studied factors sustaining environmental protection activities in communities. The results showed the longer an intervention continued, the more likely its goals were to be achieved, given the support of a residents' association equipped to deal with specific issues.

In Myojin, for example, the electric fences were installed in 2008 with assistance from the prefecture. However, for these fences to remain effective, it is necessary to cut the surrounding grass and to inspect the voltage. Hence, electric fence maintenance was set as a new goal, and the locally organized Wild Boar Countermeasure Committee aggressively supported these activities. However, in Momurahonden while interventions such as cutting grass, preparing maps, and cutting down unnecessary fruit trees were similarly implemented, the situation within the district was fluid, with no organization specializing in human-wildlife conflict interventions. The chairman of the residents' association was also replaced every year, leading to a lack of continuity.

An intervention may also be sustained via active guidance from agricultural extension advisors from regional rural development offices. A total of 45 people (as of March 2012) graduated the Satoyama Wildlife Management Technologist Training Program – a program conducted jointly by the Satoyama Science Research Center (currently the Center for Weed and Wildlife Management) of the Faculty of Agriculture of Utsunomiya University and Tochigi Prefecture (Utsunomiya University 2012). These graduates now work both inside and outside of Tochigi Prefecture. As experts on wildlife countermeasures, they are expected to participate actively in interventions within their respective regions.

5.1.4.2 Broad Participation of Residents

The second challenge facing the Model District Program is promoting wildlife awareness among all residents. The idea is that both famers and nonfarmers should participate in the interventions. In a district with many households, it is difficult to make information universally available. Only some of these households have experience of human-wildlife conflicts, which presents a challenge to awareness raising about wildlife countermeasures. In Myojin, for instance, there is a gap in both the awareness and the degree of active involvement in wildlife interventions between the west and east sides of the district. Nevertheless, damage by wild boars in the east side is on the increase. As the region is home to both long-term and new residents, it is important that all are united around wildlife issues. In this way, the west-side residents, who are well versed in human-wildlife conflict prevention measures, can share their know-how with the newcomers.

5.1.5 Conclusion: Effectiveness and Benefits of the Model District Program

Past research applied a single analysis method (a questionnaire survey) to clarify the state of human-wildlife conflict prevention interventions and residents' awareness in a specified area. However, this study employed a qualitative approach involving the observation of and active participation in activities, as well as interviews with stakeholders. The aim was to clarify the differences in residents' social backgrounds and to identify the challenges arising when the same Model District Program is implemented in different districts.

A benefit of introducing the Model District Program was opportunity creation. Specifically, the activities of the Model Program contributed by:

1. Confirming the effectiveness of wildlife countermeasures already taken (e.g., electric fences, preparing village inspection maps, etc.)
2. Providing more effective damage prevention methods
3. Providing places for residents to meet to discuss human-wildlife conflicts
4. Holding outreach and awareness-raising events for a broader range of residents

In Momurahonden, unlike in Myojin, no independent organization specializing in human-wildlife conflict interventions was in place, and residents were worried about the sustainability of the interventions. This gave rise to the following questions: For the activities to become sustainable interventions, must there be a specialist organization present to implement them? To help start a specialist organization, is it necessary to provide fixed funding for a Model Program, as was the case for the Satoyama Forest Improvement Project? Can any incentive other than funding be provided? Even in districts with specialized organizations, how will the organization and its activities be managed and sustained when the 5-year Satoyama Forest Improvement Project ends and funding stops? Thus, it is important these interventions continue to be monitored.

5.2 Testing Social-Psychological Theories to Predict Residents' Behavioral Intentions Regarding Wildlife Issues

5.2.1 Introduction

As explained in Chapter 4 and Sect. 5.1, encounters with and damage by wildlife cannot be eliminated if a limited number of residents and stakeholders are involved in wildlife management interventions. The entire community must be involved in taking measures such as managing garbage and other attractants. It should also be noted attitudes to wildlife issues are formed by various factors, including people's experience of damage and the characteristics of individual communities. Residents' attitudes may vary, even within the same district. Thus, it is often difficult for communities to share a common goal and work on interventions as one. Effective awareness-raising/education activities aimed at encouraging damage prevention interventions require an understanding of various human factors, such as people's knowledge and perceptions of wildlife issues and their willingness to engage in certain activities.

Research into the human dimensions of wildlife – a well-established academic field in the USA – utilizes social-psychological theories to help understand and predict group psychology and behavior (Decker et al. 2001). In the present context, social-psychological theories have been used to understand stakeholders and community residents' perceptions and preferences regarding wildlife and wildlife management policies. The aim is to solve conflicts between stakeholders (Baruch-Mordo et al. 2009) and to grasp and predict behavior relating to hunting and damage prevention interventions (Rossi and Armstrong 1999; Martin and McCurdy 2009).

In Japan, effective collaboration between governments and local residents is a prerequisite to successful and adaptive wildlife management. Hence, it is expedient to introduce and to utilize research relating to this human dimension. Conventional social studies in the field of wildlife management in Japan have focused either on qualitatively revealing the uniqueness and peculiarity of the relationship between people and wildlife in certain regions (Kikuchi 2006; Maruyama 2006) or quantitatively identifying residents' perceptions of damage prevention interventions and their own behavior (Kinosita et al. 2009; Yamabata 2011; Sakurai et al. 2012a). However, the only study that aimed to systematically reveal people's values and perceptions based on social-psychological theories is the empirical research conducted by Watanabe and Ogura (1996). Meanwhile, much social psychology research has been carried out into pro-environmental behaviors. General models regarding determinants that affect behavior (Hirose 1994) and models addressing the transition from perception of environmental issues to behavior (Misaka 2003) have been proposed and are subject to empirical research (Suwa et al. 2006;

Hayabuchi 2008). In general, a model is used to predict phenomena. A generalized model-based prediction is referred to as a theory (Mizuhara 1989). Using social-psychological theories, it is possible to organize diverse factors (e.g., people's perceptions and behavior), build new predictions and hypotheses (Mizuhara 1989), effectively design environmental education programs and awareness-raising campaigns, and predict the effects of these programs (Misaka 2003). Hence, a contingent benefit to wildlife management is the improved understanding of local residents and stakeholders' perceptions and behavior.

Of the numerous available social-psychological theories, I focused here on two that are related closely to interventions aimed at preventing damage caused by wildlife. I used questionnaire surveys to test hypotheses derived from these theories. The aim was to clarify both the effectiveness of these theories and to identify additional factors involved in predicting people's willingness to engage in damage prevention interventions. Based on these results, I put forward some suggestions for effective outreach and awareness-raising activities in Japan.

5.2.2 *Background of the Theories Used*

The following social-psychological theories and additional factors were used to understand local residents' perceptions.

5.2.2.1 The Theory of Planned Behavior

The Theory of Reasoned Action is one of the most commonly used social-psychological theories by which to infer people's perceptions and behavior in the field of natural resource management (Ajzen and Fishbein 1980). This theory assumes that a rational person considers the possible outcomes before acting. It helps predict people's behavioral intention by grasping their attitudes toward behavior and subjective norms (Ajzen and Fishbein 1980). "Attitudes toward behavior" refers to perceptions that determine whether a behavior is good or bad (e.g., when going to the supermarket, it is good to bring a shopping bag and not use plastic bags). A subjective norm refers to other people's expectations and the extent to which an individual modifies their behavior to meet these expectations (e.g., an important figure indicates that I should bring a shopping bag and not use plastic bags). First put forward in 1967, this theory has been verified by many case studies. Subsequently, perceived behavioral control – whether it is easy to perform the behavior (e.g., bring a shopping bag) – emerged as an important predictive factor. The Theory of Planned Behavior has since become widely used as an improved version of the Theory of Reasoned Action (Ajzen 1985).

The Theory of Reasoned Action and the Theory of Planned Behavior have been frequently verified in the field of wildlife management in North America (e.g., intention to participate in hunting (Rossi and Armstrong 1999), etiquette of backpackers in a national park (Martin and McCurdy 2009), and landowners' intention to participate in a program to protect endangered species (Sorice and Conner 2010)). In Japan, only a limited number of papers (e.g., Hayabuchi 2008; Sudo 2008) have addressed these theories, and none in the field of wildlife management.

5.2.2.2 Wildlife Acceptance Capacity Model

First advocated in North America in the late 1980s (Gigliotti et al. 2000), the concept of wildlife acceptance capacity is another theory used in this survey. When creating a population dynamic model or managing a population, the biological carrying capacity is often used to determine the maximum animal population that particular resources can support (Krebs 2001). In addition, in the 1980s, a similar concept known as "cultural carrying capacity" was proposed to determine the maximum amount of wildlife with which people can coexist in a given environment (Carpenter et al. 2000). Subsequently, Decker and Purdy (1988) defined wildlife acceptance capacity as "the maximum wildlife population that people can accept in an environment." Wildlife acceptance capacity is determined by people's perceptions of the costs/benefits linked to the presence of wildlife. Riley and Decker (2000) demonstrated that anxiety and concern (risk perception), and attitudes toward cougars (*Puma concolor*), affected the acceptance capacity of residents in Montana, USA, for cougars. Similarly, Majic et al. (2011) revealed that people's acceptance capacity toward brown bears (*Ursus arctos*) decreased when management policies were changed to protect these bears in Croatia.

When local government officials design and implement a wildlife management policy to meet the needs of residents, it is important that they understand the community's maximum acceptable wildlife population (Riley and Decker 2000; Zinn et al. 2000). However, only two studies have been conducted to measure wildlife acceptance capacity in Japan: one by Akiba et al. (2012) on raccoons (*Procyon lotor*) and the other by Sakurai et al. (2013b) on Asiatic black bears. Meanwhile, in North America a significant amount of research has been conducted on the acceptance capacity for both the wildlife populations and specific management policies (e.g., eradication and relocation).

5.2.2.3 Additional Factors

Additional factors that I considered related to community characteristics and people's perceptions of wildlife. I analyzed the perceptions of government performance, the seriousness of community issues, socio-demographics, and experience of damage by communities and individuals. Previous research has demonstrated that people's willingness to engage in damage prevention interventions depends on local governments explaining the mechanisms involved to residents, who in turn must have the opportunity to express their opinions and needs (Sakurai et al. 2013b). For communities faced with depopulation and aging, it is challenging for residents to work on interventions under their own initiative (The Nature Conservation Society of Japan 2010). As explained previously (see Sects. 4.3 and 5.1), interviews with local residents revealed how difficult it is to promote community-based damage prevention interventions. Perceptions of government performance and the seriousness of community issues – important factors likely to affect both local residents' behavioral intention to engage in damage prevention interventions and their wildlife acceptance capacity – were included in the analysis. Previous studies have also shown that both the willingness of individuals to perform certain behaviors and their wildlife acceptance capacity are affected by socio-demographics and experience of damage (Roskaft et al. 2003; Ericsson and Heberlein 2003; Agee and Miller 2009; Hirose et al. 2010).

This research thus addresses the following two points:

- The extent to which the three factors of the Theory of Planned Behavior (attitudes toward behavior, subjective norm, perceived behavioral control) and four additional factors (perceptions of government performance; seriousness of issues such as community depopulation and aging; socio-demographics such as gender, age, and occupation; experience of damage by communities and individuals) explain people's behavioral intention to engage in damage prevention interventions
- The extent to which anxiety, concern, and attitudes toward wildlife and four additional factors (perceptions of government performance, seriousness of community issues, socio-demographics, experience of damage by communities and individuals) explain people's wildlife acceptance capacity

5.2.3 Study Sites

Surveys were conducted in the Myojin district of Nikko City and the Fukahodo district of Kanuma City, Tochigi Prefecture (Fig. 5.1, Table 5.1). In 2007, Tochigi was ranked third in Japan for agricultural damage caused by wildlife (7,714 tons) (Ministry of Agriculture, Forestry and Fisheries 2007).

Located in the northwestern part of the prefecture, Nikko City has suffered significant wildlife damage. In FY2010, the amount of agricultural damage was 17,541,000 yen (about USD 175,410) (personal communication with a staff member at Nikko City Government). Located in the southern part of Nikko City, Myojin has 538 households, with an average age of 49.5 (population aging rate, 32.1%) (Nikko City 2010). Agricultural damage is caused by wildlife such as wild boars and macaques, primarily in the western part of the district that lies close to the mountains. The Tobu Nikko Railway runs through the eastern part of the district, and residential areas have been developed around Myojin Station. Encounters with and damage caused by wild boars have increased since 2011.

In 2008, as explained previously (see Sect. 5.1), electric fences were erected close to the mountains in the western part of the Myojin district. In 2009, the

Fig. 5.4 Landscape of Fukahodo. (Photo taken by the author)

Satoyama Woodland Improvement Program was introduced to reduce wildlife damage via the Prefectural Tax for Healthy Forest Creation. Similarly, wildlife damage prevention interventions were implemented for a period of 4 years using grants from the prefectural government (e.g., removal of underbrush). Continuous efforts included the residents checking the voltage of the electric fences and clearing weeds around them. The residents of the Myojin district are socio-demographically diverse. Some actively engage in damage prevention interventions with the support of the local government, while others do not. The native residents are mostly farmers, while the new residents living near to the station in the eastern part of the district are mostly the families of office workers.

Fukahodo (Fig. 5.4), with 264 households, is located in the southwestern part of Kanuma City. The average age in the Kiyosu district (including Fukahodo and neighboring districts) is 46.6 (population aging rate, 24.2%) (Kanuma City 2010). Wild boars and deer are the two species that cause the most serious damage. Agricultural damage in FY2008 amounted to 14,510,000 yen (about USD 145,100) (Kanuma City 2008).

In Fukahodo, wild boars caused significant damage to agricultural land close to the mountains in the western part of the district. As in the case of Myojin, Fukahodo introduced the Satoyama Woodland Improvement Program utilizing the Prefectural Tax for Healthy Forest Creation in FY2010. In the same year, Fukahodo was designated a model district to reduce human-wildlife conflicts. The Fukahodo Satoyama

Protection Association was organized primarily by the residents' association to promote wildlife damage prevention interventions. Thus, Myojin and Fukahodo have similar characteristics. As in the case of Myojin, Fukahodo needed to raise awareness about wildlife across the district, especially among nonfarmers (Sakurai 2019). In Myojin and Fukahodo, where residents are socio-demographically diverse, it is important to understand the factors motivating people to engage in interventions.

5.2.4 Details of the Questionnaire Survey

A questionnaire survey was conducted in Myojin and Fukahodo, with cooperation from the respective residents' associations. In September 2011, a questionnaire was distributed to all households (Myojin, $n = 315$; Fukahodo, $n = 189$) affiliated with the residents' associations. One month later, the completed questionnaires were collected by the group leaders of the respective districts. The respondents answered anonymously.

The questionnaire consisted of seven items: four for measuring the Theory of Planned Behavior (*independent variables*, 1.1 attitudes toward the interventions, 1.2 subjective norm, 1.3 perceived behavioral control on the interventions; *dependent variable*, 1.4 behavioral intention to engage in damage prevention interventions) and three factors for measuring wildlife acceptance capacity (*independent variables*, 2.1 anxiety and concern toward wildlife, 2.2 attitudes toward wildlife; *dependent variable*, 2.3 wildlife acceptance capacity) (Table 5.2). To determine the wildlife acceptance capacity, the respondents were asked about wild boars, which caused the greatest damage in both districts.

When measuring the three elements of the Theory of Planned Behavior (attitudes toward interventions, subjective norm, perceived behavioral control on the interventions), the use of multiple questionnaire items increases the reliability of each element and is generally recommended (Pouta and Rekola 2001; Hayabuchi 2008). While measurements based on multiple items help verify the theory, however, there is a resultant increase in the number of questions. However, as my priority in this research was to reduce the burden on respondents, my verification of the theory was based on a simplified method. Specifically, I used only one item each for the attitudes toward the interventions and perceived behavioral control and two items for the subjective norm (viewpoint of neighbors and family members).

Regarding the model's additional factors, I used five items for the perceptions of government performance (3.1) and four items for the seriousness of community issues (4.1) (Table 5.2). For socio-demographics, I used gender (1 = female, 2 = male), age (1 = 20s or younger, 2 = 30s, 3 = 40s, 4 = 50s, 5 = 60s, 6 = 70s or older), ownership of agricultural land (1 = yes, 0 = no), and experience of damage (1 = yes, 0 = no) as additional factors. From July 2011, I conducted interviews with residents about the current status of wildlife damage and interventions in Myojin and Fukahodo. Participant observations at seminars and community field trips aimed at promoting interventions preventing damage by wild boars were also noted.

Table 5.2 Questionnaire items and descriptive results

Major category of questions	Subcategory of questions	Answer scale	Mean score (n) Myojin/Fukahodo	Standard deviation Myojin/ Fukahodo	Cronbach alpha's score Myojin/ Fukahodo
1.1 Attitudes toward interventions	It is good to engage in wildlife damage prevention interventions	Five-point scale (1. disagree, 2. slightly disagree, 3. neither agree nor disagree, 4. slightly agree, 5. agree)	4.34(196)/4.55(96)	0.96/0.74	
1.2 Subjective norm	Neighbors expect me to engage in wildlife damage prevention interventions		2.39(190)/2.55(94)	1.45/1.37	0.86/0.73
	Family members expect me to engage in wildlife damage prevention interventions		2.61(190)/2.82(97)	1.47/1.48	
1.3 Perceived behavioral control of the interventions	I know how to prevent damage by wildlife		2.35(185)/2.66(90)	1.35/1.43	
1.4 Behavioral intention regarding damage prevention interventions	I am willing to engage in wildlife damage prevention interventions		3.45(190)/3.80(92)	1.28/1.05	
2.1 Anxiety and concern toward wildlife	Wildlife-human conflicts have been increasing recently		4.42(194)/4.55(99)	0.97/0.88	0.66/0.76
	I am concerned about crop damage caused by wildlife that lives in and around the district		4.06(198)/4.57(97)	1.15/0.79	

(continued)

Table 5.2 (continued)

Major category of questions	Subcategory of questions	Answer scale	Mean score (n) Myojin/Fukahodo	Standard deviation Myojin/ Fukahodo	Cronbach alpha's score Myojin/ Fukahodo
2.2 Attitudes toward wildlife	Wildlife helps maintain a healthy natural environment		3.28(194)/2.98(96)	1.41/1.31	
2.3 Wildlife acceptance capacity	Population of wild boars in and around the district	Three-choice question (1. should be reduced, 2. should remain unchanged/ have no idea, 3. should be increased)	1.45(199)/1.18(92)	0.53/0.42	
3.1 Perceptions of government performance	The city government has properly listened to residents' opinions about wildlife issues	Five-point scale (1. disagree, 2. slightly disagree, 3. neither agree nor disagree, 4. slightly agree, 5. agree)	3.13(193)/3.08(95)	1.19/1.30	0.86/0.88
	The prefectural government has properly listened to residents' opinions about wildlife issues		3.03(190)/2.85(97)	1.19/1.24	
	The city government has provided residents with sufficient information about how to cope with wildlife issues		3.07(193)/2.87(95)	1.17/1.29	
	The prefectural government has provided residents with sufficient information about how to cope with wildlife issues		2.90(195)/2.74(96)	1.08/1.23	

(continued)

Table 5.2 (continued)

Major category of questions	Subcategory of questions	Answer scale	Mean score (n) Myojin/Fukahodo	Standard deviation Myojin/Fukahodo	Cronbach alpha's score Myojin/Fukahodo
	The information about wildlife provided by local governments is reliable		3.55(195)/3.55(96)	1.13/1.07	
4.1 Seriousness of community issues	Decreased labor force in the community: aging, lack of young people, etc.	Five-point scale (1. not serious, 2. not very serious, 3. not sure, 4. slightly serious, 5. serious)	4.05(193)/4.10(89)	0.93/0.81	0.81/0.83
	Public transport issues in the community, issues related to hospitals and nursing care facilities, etc.		3.84(176)/3.98(85)	0.92/0.90	
	Livelihood issues in the community: increase in abandoned farmland, difficulty in conducting funerals and festivals due to lack of manpower, etc.		3.27(175)/3.91(85)	1.03/0.93	
	Pressure from outside the community: wildlife damage, typhoon damage, etc.		3.42(171)/3.92(88)	1.07/0.81	

Based on Sakurai et al. (2014a)

These interviews revealed various findings about people's attitudes toward wildlife issues within each community (Sakurai et al. 2013a), and the questionnaire was constructed based on this information.

5.2.5 *Analysis*

To analyze the results, I removed responses with no answers from those with valid answers. For factors which consisted of two or more items (subjective norm, anxiety and concern toward wildlife, perceptions of government performance, seriousness of community issues), the average value (average value of answers based on the five-point scale) was used as the score for analysis. Cronbach's alpha score was calculated to evaluate internal consistency. To verify the theory, I conducted a multiple regression analysis to assess the correlations between items. Percentage variance was expressed via R^2 to indicate the explanatory power of the model. To analyze the measurements of the Theory of Planned Behavior, I used the attitudes toward the interventions, subjective norm, and perceived behavioral control as independent variables. Behavioral intention toward the damage control interventions was the dependent variable. For the wildlife acceptance capacity model, I used anxiety and concern toward wildlife and attitudes toward wildlife as the independent variables. Wildlife acceptance capacity was the dependent variable. To investigate the independent variables' multicollinearity, I measured the variance inflation factor (VIF) and the condition index. In general, if the VIF is 2 or less, or if the condition index is less than 30, the level of multicollinearity is considered low (Murase et al. 2009). A p value of 0.05 was considered significant in all cases. SPSS 2010 (IBM, Tokyo, Japan) software was used for the statistical analysis.

5.2.6 *Results*

5.2.6.1 Descriptive Statistics

In Myojin, there were 219 responses for 315 questionnaires distributed (response rate, 69.5%). In Fukahodo, there were 99 responses for 189 questionnaires distributed (response rate, 52.4%). The percentage of male respondents was high in both districts (Myojin, 69.7%; Fukahodo, 75.3%). The average age of the respondents was 56 in both Myojin and Fukahodo, which was slightly higher than the average age in these districts. In terms of occupation, the unemployed accounted for the largest portion of the respondents in Myojin (33.8%), followed by company employees (29.2%) and agriculture and forestry workers (12.8%). In Fukahodo, company employees accounted for the largest portion (42.9%), followed by the unemployed (24.2%) and housewives (14.3%). In Myojin, agricultural land owners accounted for 36.1% of the respondents, while 35.0% of respondents had experience of wildlife damage. In Fukahodo, agricultural land owners accounted for 64.8% of the respondents, while 56.3% had experience of wildlife damage.

Residents in both districts had favorable attitudes toward the interventions (Question 1.1, Myojin, average value = 4.34; Fukahodo, average value = 4.55). Meanwhile, the subjective norm (Question 1.2, Myojin, average value = 2.50; Fukahodo, average value = 2.73) and perceived behavioral control (Question 1.3, Myojin, average value = 2.35; Fukahodo, average value = 2.66) were low. The average values of behavioral intention to engage in damage prevention interventions (Question 1.4) were 3.45 in Myojin and 3.80 in Fukahodo (Table 5.2).

Residents had high levels of anxiety and concern toward wildlife (Question 2.1, Myojin, average value = 4.24; Fukahodo, average value = 4.56), and attitudes toward wildlife (Question 2.2) were positive and negative in the same proportion (Myojin, average value = 3.28; Fukahodo, average value = 2.98). Regarding the population of wild boars around the district (Question 2.3), 56.8% of the respondents in Myojin replied that the population should be reduced, and 38.2% replied that they had no idea. In Fukahodo, 82.6% of the respondents believed that the population should be reduced, while 14.1% stated that they had no idea (Table 5.2).

Regarding perceptions of government performance (Question 3.1), the item "The information offered by local governments is reliable" scored highest in both districts (Myojin, average value = 3.55; Fukahodo, average value = 3.55). The average values of the five items were 3.15 in Myojin and 3.02 in Fukahodo. Regarding the seriousness of community issues (Question 4.1), the item "Decreased labor force in the community" scored the highest (Myojin, average value = 4.05; Fukahodo, average value = 4.10), and the average values of the four items were 3.72 in Myojin and 3.99 in Fukahodo (Table 5.2).

Reliability analysis gave a Cronbach's alpha score of 0.6 or more (Table 5.2) in both of the districts. Therefore, we judged the internal consistency of the respective items to be acceptable.

5.2.6.2 Correlations Among the Items

Multiple regression analysis revealed that the attitudes toward the interventions ($B = 0.23$, $P < 0.01$), subjective norm ($B = 0.38$, $P < 0.01$), and experience of damage ($B = 0.17$, $P < 0.05$) showed a significant correlation with behavioral intention to engage in damage prevention interventions in Myojin (Table 5.3). The percentage variance explained by the model was 45.2%. The collinearity test revealed that all VIFs were 2 or less, and the correlation among the items was low (Murase et al. 2009).

In Fukahodo, the subjective norm ($B = 0.33$, $P < 0.01$), perceived behavioral control of the interventions ($B = 0.26$, $P < 0.05$), and experience of damage ($B = 0.31$, $P < 0.01$) showed a significant correlation with the behavioral intention to engage in damage prevention interventions. The percentage variance explained by the model was 42.0%. The collinearity test showed that all VIFs were 2 or less, and the correlation among the items was low (Table 5.3).

Table 5.3 Factors explaining behavioral intention to engage in wildlife damage prevention interventions in Myojin (n = 160) and Fukahodo (n = 80)

	Independent variables	*B*: standardized coefficients	*r*: correlation	*VIF*	Condition index	R^2	Adjusted R^2
Myojin	Attitudes toward the interventions*	0.23**	0.42**	1.22	25.93	0.48	0.45
	Subjective norm*	0.38**	0.57**	1.45			
	Perceived behavioral control of the interventions*	0.05	0.34**	1.36			
	Perceptions of government performance	0.11	0.25**	1.20			
	Seriousness of community issues	0.03	0.32**	1.27			
	Gender	0.03	0.21**	1.21			
	Age	-0.04	0.17***	1.16			
	Ownership of agricultural land	0.10	0.43**	1.93			
	Experience of damage	0.17***	0.44**	1.97			
Fukahodo	Attitudes toward the interventions*	0.15	0.21***	1.08	36.34	0.49	0.42
	Subjective norm*	0.33**	0.52**	1.45			
	Perceived behavioral control of the interventions*	0.26***	0.52**	1.55			
	Perceptions of government performance	0.00	0.21***	1.24			
	Seriousness of community issues	-0.03	0.20***	1.25			
	Gender	0.10	0.29**	1.11			
	Age	-0.05	0.07	1.06			
	Ownership of agricultural land	-0.05	0.25***	1.40			
	Experience of damage	0.31**	0.45**	1.47			

Based on Sakurai et al. (2014a)
*Factors for measuring the Theory of Planned Behavior. ** $p<0.01$, *** $p<0.05$

In Myojin, anxiety and concern toward wildlife ($B = -0.31$, $P < 0.01$), attitudes toward wildlife ($B = 0.24$, $P < 0.01$), and age ($B = -0.16$, $P < 0.05$) showed a significant correlation with wildlife acceptance capacity (Table 5.4). The percentage variance explained by the model was 37.4%. In the collinearity test, the VIF of the experience of damage exceeded 2 (2.07) and showed a slight collinearity. However,

Table 5.4 Factors explaining wildlife acceptance capacity in Myojin (n = 164) and Fukahodo (n = 80)

	Independent variables	*B*: standardized coefficients	*r*: correlation	*VIF*	Condition index	R^2	Adjusted R^2
Myojin	Anxiety and concern toward wildlife*	−0.31**	−0.50**	1.44	27.95	0.41	0.37
	Attitudes toward wildlife*	0.24**	0.42**	1.20			
	Perceptions of government performance	−0.06	−0.13	1.09			
	Seriousness of community issues	−0.04	−0.26**	1.26			
	Gender	−0.11	−0.26**	1.21			
	Age	−0.16***	−0.31***	1.19			
	Ownership of agricultural land	0.00	−0.34**	1.95			
	Experience of damage	−0.14	−0.41**	2.07			
Fukahodo	Anxiety and concern toward wildlife*	−0.17	−0.30**	1.33	34.52	0.23	0.15
	Attitudes toward wildlife*	0.17	0.17	1.10			
	Perceptions of government performance	0.15	0.08	1.05			
	Seriousness of community issues	−0.28***	−0.32**	1.18			
	Gender	−0.14	−0.15	1.13			
	Age	0.01	−0.03	1.15			
	Ownership of agricultural land	−0.19	−0.21***	1.36			
	Experience of damage	0.17	−0.07	1.43			

Based on Sakurai et al. (2014a)
*Factors for measuring wildlife acceptance capacity. **$p<0.01$, ***$p<0.05$

the condition index was less than 30 (27.95). I, therefore, concluded this was sufficient for the analysis.

In Fukahodo, only the seriousness of community issues ($B = -0.28$, $P = 0.05$) showed a significant correlation with the wildlife acceptance capacity (Table 5.4). The percentage variance explained by the model was 14.7%. In the collinearity test, all the VIFs were below 2, and the correlation among items was low (Table 5.4).

5.2.7 Discussion

Regarding the Theory of Planned Behavior, two factors showed a significant correlation with the residents' behavioral intention in both districts. For the wildlife acceptance capacity model, two factors showed a significant correlation in Myojin. This indicates that the theories and models developed in the USA are somewhat effective in predicting people's behavioral intention and perceptions in Japan – a country with different cultural and social backgrounds. Thus, the theories have external validity (i.e., similar results may be obtained in different groups and situations). The fact that perception structures showed a similar tendency across communities and cultures suggests that worldwide case studies could be applied in different countries, including Japan, to implement awareness-raising campaigns and to encourage damage prevention interventions.

It should be noted that not all factors displayed explanatory significance. The details and extent of explanatory factors varied according to the district, and additional factors such as experience of damage and socio-demographics had significant explanatory power. To verify people's behavior and perceptions accurately, it hence is essential to understand the social background and history of a community, as well as the factors underpinning the existing theories.

5.2.7.1 Verification of the Theory of Planned Behavior

In the previous studies related to the Theory of Planned Behavior in North America, the explanatory power of the subjective norm was low for factors affecting interventions (e.g., preventing damage by brown bears and people's intention to participate in hunting) (Martin and McCurdy 2009; Hrubes et al. 2001). In this research, the subjective norm affected the behavioral intention significantly in both Myojin and Fukahodo. This supports the results of previous studies, which concluded that, unlike in Europe and the USA, the subjective norm is an important element explaining behavior in Japan – a country where people find it difficult to ignore their perception by others (Ohtomo 2004; Hashimoto et al. 2008). In both districts, the more residents perceived that family members and neighbors expected them to work on damage prevention interventions, the more they were motivated to do so. Thus, it is reasonable to assume that encouragement by others is important in promoting interventions by residents and that this encouragement is facilitated when the need for damage prevention interventions is recognized throughout the district. Consequently, a lack of shared recognition may nullify this same encouragement. Similarly, encouragement may cause problems between residents whose recognition of the need for action is different. Thus, awareness-raising campaigns should be organized to ensure the mutual recognition of damage prevention interventions across the district.

In addition to the subjective norm, the residents' behavioral intention in Myojin was affected by the attitudes toward the interventions. In Fukahodo, meanwhile, it was affected by the perceived behavioral control of the interventions. The two districts have a lot in common, including damage caused by wild boars, mainly in areas close to the mountains. However, the programs implemented to reduce human-wildlife conflicts were different. In Myojin, residents engaged in an ongoing process of weeding and electric fence inspection in the western area close to the mountains. In FY2008, most of the local households participated in erecting a 1.8 km stretch of electric fencing (Sakurai et al. 2013a). The ongoing and participatory program to reduce human-wildlife conflicts in the western part of Myojin may have affected the residents' attitudes toward both the interventions and the norm, which increased the correlation between these variables and the behavioral intention. In Fukahodo, in 2011, mountainside weeding was undertaken by contractors, and only certain residents (such as farmers) participated in erecting electric fences. To increase awareness about wildlife damage across the district (including among nonfarmers), a family education program regarding interventions to prevent damage by wild boars was held in Fukahodo about 2 months before the distribution of the questionnaire survey. The event was organized as part of the prefecture's Model District Program. Its aim was to provide information and knowledge about preventing damage by wild boars and to explain the importance of the interventions (Sakurai 2019). About 80 people (approximately one-third of the total households in the district) took part. This may have helped increase the value of perceived behavioral control of methods undertaken to prevent damage in Fukahodo, compared to Myojin. It may also have enhanced the correlation between perceived behavioral control and behavioral intention.

Experience of damage was found to have affected behavioral intention in both districts. In Myojin and Fukahodo, the area in which residents encountered wild boar has expanded in recent years. Those residents yet to suffer damage are likely to do so in the future. Hence, it is important to provide information about wildlife issues within the community. Measures to achieve this may include a circular newsletter or map that clearly indicates zones within the district likely to suffer damage (Fig. 5.5), all of which may help increase awareness of wildlife damage among residents who are yet to experience it themselves (Enari and Suzuki 2010).

5.2.7.2 Verification of the Wildlife Acceptance Capacity Model

In both districts, most of the respondents thought that the population of wild boars should be reduced. Thus, the wild boar population around these districts was deemed to have exceeded both the cultural carrying and acceptance capacity. In Myojin, only 35% of respondents had suffered damage by wild boars, and not many had spotted wild boars in the district. However, indirect experience (e.g., hearing about damage from neighbors) is likely to have increased anxiety (Jonker et al. 2009). According to Majic et al. (2011), the increased population of brown bears in Croatia resulted in a significant decrease in people's acceptance capacity. In Myojin and Fukahodo, the increased experience of damage and rise in encounters over the past 10 years are likely to have raised residents' concern about unknown risk (resulting

Fig. 5.5 A map indicating zones where wildlife damage could occur was created through collaborative efforts involving farmers, researchers, local government officials, and other stakeholders. The photo was taken by the author in one of the model districts (Ogaki district, Tochigi City)

from passive and unfamiliar experience), leading to a low acceptance capacity toward wild boars.

In Myojin, the acceptance capacity was affected by factors including attitudes toward wildlife, age, anxiety, and concern. As stated above, the district has two distinct communities: one in the western part, close to the mountains, which is populated mainly by farmers and senior citizens, and one in the eastern part, around the Myojin Station of the Tobu Railway, which is populated by relatively young families, most of whom are office workers. Residents living around the station have little experience of wildlife damage. Hence, they may have positive attitudes toward wildlife and a high acceptance capacity. Senior citizens who have suffered damage, meanwhile, may have a low acceptance capacity (Sakurai et al. 2014c). If this is indeed the case, awareness-raising campaigns for senior residents in Myojin may increase their acceptance capacity. During the interviews, I was often told by residents that all the wild boars in the vicinity should be captured and destroyed. However, as this would be difficult to achieve both in practice and politically, local residents tolerate a certain population of wild boars. Thus, it is necessary to disseminate knowledge and skills to promote damage prevention interventions and to provide information about positive elements, such as the value of wildlife (e.g., boars as symbols of a rich natural environment).

In Fukahodo, the wildlife acceptance capacity was affected by the seriousness of community issues. The values for all the items relating to community issues were higher than for Myojin, suggesting a sense of crisis among residents in regard to their community. The more seriously residents perceive livelihood issues (e.g., a decreased labor force due to a lack of young people and increased abandoned farmland), the lower the wildlife acceptance capacity. This result is in line with those of the preliminary interview and previous research (The Nature Conservation Society of Japan 2010). Wildlife issues are an integral part of community life, and measures should be taken to re-empower communities (e.g., promoting agriculture) to increase residents' wildlife acceptance capacity and to build resilience against wildlife damage.

5.2.7.3 Limitations of the Research

A limitation of this research is that behavioral intention was measured using abstract expressions (Question 1.4 "I am willing to engage in wildlife damage prevention interventions" in Table 5.2). Therefore, the respondents may have had various different interventions in mind while answering (e.g., erecting electric fences, removing underbrush around the community). In social psychology research into pro-environmental behavior, the intention of specific behavior is often measured to avoid this issue (Yorifuji and Hirose 2002; Ohtomo 2004). Future studies could incorporate more specific behaviors and consider factors such as intention, attitudes, subjective norm, and perceived behavioral control. This would help clarify the specific causes of the differences in correlation among items in the respective communities.

5.2.7.4 Suggestions for Potential Awareness-Raising Activities

Previous research found that by promoting techniques to prevent damage and reducing damage through community-wide efforts, residents' confidence in conducting wildlife damage prevention intervention can be increased (Yamabata 2010). Other studies revealed that seminars organized by local governments can help increase residents' confidence and involvement in such interventions (Sakurai et al. 2012b, 2013b). In both Myojin and Fukahodo, unique, ongoing awareness-raising activities should be organized for specific targets, in addition to a general district-wide outreach program. This will help build a community where residents have both knowledge about wildlife damage prevention interventions and a certain level of wildlife acceptance capacity. The survey results showed that the determinants of residents' behavioral intention and perceptions were different for the two communities. Thus, measures should be implemented that are appropriate to the specific characteristics of each district. In Myojin, for example, efforts should be made to increase the awareness of senior residents whose wildlife acceptance capacity is low.

The survey results offered also basic information allowing local governments to shift from conventional reactive interventions (e.g., interventions in response to damage, awareness-raising activities in response to residents' requests) to proactive (e.g., preliminary interventions to provide information in communities where the acceptance capacity is low and residents are highly dissatisfied and anxious) and adaptive (e.g., interventions designed according to district characteristics and changes in residents' perceptions) interventions. Continuous monitoring will be necessary to see how the current awareness-raising activities (e.g., a series of seminars and community field trips as part of the program to reduce human-wildlife conflicts) change the residents' norm, attitudes toward interventions, confidence, and wildlife acceptance capacity. The extent to which such activities help increase the number of residents involved in damage prevention interventions could also be assessed.

5.3 Residents' Perceptions of Wildlife Conflicts in Momurahonden, Nasushiobara City: Comparison with the Previous Year and Neighboring Communities

5.3.1 Research Background and Process

I conducted a survey in the Momurahonden district (number of households, 57) located in the northeastern part of Nasushiobara City, Tochigi Prefecture (Fig. 5.1, Table 5.1). As explained in Sect. 5.1, macaques, black bears, and deer frequently appeared in and around the residential area, and local residents had suffered agricultural damage. More recently, wild boars had also started to appear. To mitigate the human-wildlife conflicts, Momurahonden was chosen as a target district under the Satoyama Woodland Improvement Project in 2010. In the same year, the district was also designated a model district under the program organized by the Nature Preservation Division of the prefectural government. Under these programs, local residents participated in removing weeds, conducting community inspections, clearing abandoned bamboo thickets, and holding seminars about ways to scare away macaques.

This chapter reports the results of a questionnaire survey conducted in 2011. The aim was to understand the residents' perceptions of wildlife damage, as well as to compare results with (1) those of the questionnaire survey conducted in the same district in 2010 and (2) those of a questionnaire survey conducted in 2011 in neighboring districts where the Model District Program and other similar programs were not implemented. My discussion of the potential effectiveness of the Satoyama Woodland Improvement Program and the Model District Program is based on these results.

5.3.2 Method

In November 2011, I conducted a questionnaire survey for all households (n = 57) affiliated with the Momurahonden residents' association. The questionnaires were distributed and collected with the help of the association, and a representative of each household was asked to answer the survey. A questionnaire survey for the previous year (May 2010) was conducted by the Tochigi Prefectural Forest Management Office, and the results used for comparison. Questionnaires were distributed to all households in the district in a similar manner. For comparison, a questionnaire survey was conducted in March 2012 in Shigiuchi and Yugu (adjacent to Momurahonden) – districts where local governments had not actively implemented wildlife interventions such as those set out in the Model District Program. The number of households in Shigiuchi and Yugu were 90 and 41, respectively (Nasushiobara City 2010). The questionnaires were sent by mail to all of the households listed in the telephone directory (Shigiuchi, 65; Yugu, 38) (NTT East Japan 2011).

Table 5.5 shows the questionnaire items. The results were compiled for comparison by districts over a 2-year period. For perceptions of government performance and entities that should work on interventions, I used the Mann-Whitney U test to determine the difference in mean value by district. I also used the Mann-Whitney U test to determine the difference in socio-demographics and perceptions between residents who continued to participate in the Model District Program and those who no longer did so. SPSS18 (IBM, Tokyo, Japan) software was used for the analysis, and the significance level was set at $p = 0.05$.

5.3.3 Results and Discussion

In the 2011 Momurahonden survey, questionnaires were collected from 24 households (response rate, 42%). Most of the respondents were male (72%) in their 60s or older (60s, 28%; 70s, 33%). In the 2010 survey conducted by the Tochigi Prefectural Forest Management Office, questionnaires were collected from 40 households. In the 2011 survey of neighboring districts (Shigiuchi and Yugu), 47 households responded (response rate, 46%). Almost all the respondents were male (91%). In terms of age, the respondents in their 60s accounted for the largest proportion (39%).

In the 2011 Momurahonden survey, 13% of the respondents spotted macaques more than 30 times and 52% between 10 and 29 times in the past year (Table 5.6). The results of the survey remained almost unchanged from the previous year (10% and 53%, respectively). The percentage of respondents suffering damage by macaques remained almost unchanged (54% in 2011 and 57% in 2010). The same was true for the percentage of respondents who worked on the interventions against macaques (64% in 2011 and 61% in 2010) (Table 5.7). These results show that residents continued to encounter macaques and suffered damage from 2010 to 2011

Table 5.5 Questions and answer scale

Major category of questions	Subcategory of questions	Answer scale
Socio-demographics	Gender	1. male, 2. female
	Age	1. 10s, 2. 20s, 3. 30s, 4. 40s, 5. 50s, 6. 60s, 7. 70s or older
Macaque sightings, damage, and status of interventions	How often did you spot macaques in the past year?	1. none, 2. less than 10 times, 3. 10 to 29 times, 4. more than 30 times
	Did you suffer damage from macaques in the past year?	1. yes, 2. no
	Do you engage in interventions against macaques?	1. yes, 2. no
Cultivation of crops and frequency of weeding around the farmland	Do you cultivate farmland crops?	1. yes, 2. no
	How often do you remove weeds from the farmland?	1. seldom, 2. about once a year, 3. about 2–4 times a year, 4. 5 times or more a year
Status of wildlife damage and seriousness	Have you suffered wildlife damage in the past year?	1. yes, 2. no
	Was the wildlife damage serious?	1. disagree, 2. slightly disagree, 3. neither agree nor disagree, 4. slightly agree, 5. agree
Status and details of wildlife interventions	Do you currently engage in wildlife damage prevention interventions?	1. yes, 2. no
	Which specific wildlife damage prevention interventions are you engaged with?	1. weeding, clearing of thickets; 2. garbage management; 3. electric fences; 4. wire mesh fences; 5. protective nets; 6. cutting back unnecessary fruit trees; 7. scaring wildlife away using fireworks, etc.; 8. others
Perceptions of government performance regarding wildlife issues	The city government has properly listened to residents' opinions about wildlife issues	1. disagree, 2. slightly disagree, 3. neither agree nor disagree, 4. slightly agree, 5. agree
	The prefectural government has properly listened to residents' opinions about wildlife issues	
Entity that should work on wildlife interventions	Farmers who suffered damage	
	District residents	
	Local governments	

(continued)

Table 5.5 (continued)

Major category of questions	Subcategory of questions	Answer scale
Participation in interventions as part of the Model District Program	March 2010: Community inspection	1. did not participate, 2. not sure, 3. participated
	May 2010: Seminar: interventions against macaques	
	August 2010: Cutting back unnecessary fruit trees	
	July 2011: Seminar: wild boars	
	October 2011: Cutting back unnecessary fruit trees	

Table 5.6 Two-year comparison of macaque sightings

		None (%)	Less than 10 times (%)	10~29 times (%)	More than 30 times (%)
How often did you spot macaques in the past year?	2011: Momurahonden (n = 22)	0	55	52	13
	2010: Momurahonden (n = 40)	10	28	53	10
	2011: Neighboring districts (n = 46)	9	24	24	44

Based on Sakurai et al. (2014b)

and that they continued to engage in the damage prevention interventions over the same period. In neighboring districts, 44% of respondents spotted macaques more than 30 times in the past year – a significantly higher percentage than in Momurahonden. In these districts, the percentage of respondents who worked on interventions was low (48%) (Table 5.7).

Respondents who cultivated farmland crops accounted for 77% in 2011, which was slightly lower than the previous year (88%). There are two main possible explanations for this: (1) some residents stopped cultivating crops; and/or (2) the percentage of residents who cultivated crops and did not respond to the 2011 questionnaire was high. Regarding the frequency of weeding around residential houses and farmland, 29% of the respondents did so 5 times or more a year, and 57% did so 2–4 times a year in 2011. These results were almost the same for the previous year (5 times or more a year, 30%; 2–4 times a year, 55%). The results show that the residents continued weeding despite the decrease in crop cultivation.

Regarding questions about overall wildlife damage (in addition to macaques), most of the respondents (91%) suffered wildlife damage in Momurahonden. However, only 66% considered the damage to be serious. In the neighboring districts, the percentage of respondents suffering damage was lower than that of

Table 5.7 Two-year comparison of the status of interventions

		No (%)	Yes (%)
Do you currently engage in interventions against macaques?	2011: Momurahonden (n = 22)	36	64
	2010: Momurahonden (n = 36)	39	61
	2011: Neighboring districts (n = 46)	52	48

Based on Sakurai et al. (2014b)

Momurahonden (64%). In this instance, most of them (80%) considered the damage to be serious. The percentage of respondents engaging in interventions to prevent wildlife damage in Momurahonden (59%) was higher than that of the neighboring districts (42%). The percentages of respondents who were assisting in the weeding and clearing of thickets (Momurahonden, 65%; neighboring districts, 32%) and cutting back unnecessary fruit trees (Momurahonden, 41%; neighboring districts, 18%) were significantly different in these districts. In Momurahonden, where the Satoyama Woodland Improvement Program and the Model District Program were implemented, the perceived seriousness of the damage may have been mitigated by residents' collaborative interventions.

Regarding the perceptions of government performance on wildlife issues, more than 30% of the respondents in Momurahonden chose "Agree" for two items: "The city government has properly listened to residents' opinions about wildlife issues," and "The prefectural government has properly listened to residents' opinions about wildlife issues" (Table 5.8). In the neighboring districts, 2% of residents chose "Agree" for these items. The Mann-Whitney U test showed a significant difference between these districts (perceptions of city government performance, $p < 0.01$; perceptions of prefectural government performance, $p < 0.01$). To the question of which entity should work on wildlife interventions, most of the respondents in Momurahonden chose farmers who had suffered damage (Momurahonden, 71%; neighboring districts, 52%) and residents of the district (Momurahonden, 50%; neighboring districts, 41%). Meanwhile, most of the residents in the neighboring districts chose local governments (Momurahonden, 68%; neighboring districts, 83%). There was no significant difference between these districts for this item ($p > 0.05$). In Momurahonden, in 2010, the prefectural and city governments began to hold seminars and work with residents as part of periodic interventions via the Satoyama Woodland Improvement Program and the Model District Program. These efforts were key factors in changing residents' perceptions of government's performance in Momurahonden. This result is in line with those of a previous study, which demonstrated that measures implemented by local governments improved residents' perceptions of government performance (Sakurai et al. 2012b). Other previous studies demonstrated also residents started to engage in interventions (at the prior request of local governments) under their own initiative after trust had been established between the parties (Sakurai et al. 2013b). In Momurahonden, most of the respondents thought it right that they should work on the damage prevention

Table 5.8 Perceptions of government performance and the question of which entities should work on wildlife interventions

		Disagree (%)	Slightly disagree (%)	Neither agree nor disagree (%)	Slightly agree (%)	Agree (%)
The city government has properly listened to residents' opinions about wildlife issues	Momurahonden (n = 22)	18	0	32	18	32
	Neighboring districts (n = 43)	44	23	23	7	2
The prefectural government has properly listened to residents' opinions about wildlife issues	Momurahonden (n = 23)	22	4	13	26	35
	Neighboring districts (n = 44)	50	21	23	5	2
Farmers who have suffered damage should work on interventions	Momurahonden (n = 17)	12	0	6	12	71
	Neighboring districts (n = 29)	7	0	7	35	52
Local residents should work on interventions	Momurahonden (n = 18)	11	6	11	22	50
	Neighboring districts (n = 29)	10	0	21	28	41
Local governments should work on interventions	Momurahonden (n = 19)	0	0	5	26	68
	Neighboring districts (n = 27)	0	0	0	17	83

Based on Sakurai et al. (2014b)

Table 5.9 Participation in interventions as part of the Model District Program

	Did not participate (%)	Not sure (%)	Participated (%)
March 2010: Community inspection	17	4	78
May 2010: Seminar about interventions against macaques	30	4	65
August 2010: Cutting back unnecessary fruit trees	48	4	48
July 2011: Seminar about wild boars	57	4	39
October 2011: Cutting back unnecessary fruit trees	57	0	44

Based on Sakurai et al. (2014b)

interventions under their own initiative (they chose "Farmers who suffered damage" and "Residents in the district" as the entities who ought to be engaged). A sense of responsibility and leadership may have followed on from trust being developed between local governments and residents.

Finally, the changes in the number of participants in the model district interventions in Momurahonden are presented in Table 5.9. When Momurahonden was designated as a model district, most of the respondents (78%) participated in the interventions. As time progressed, however, the number of participants decreased (about 40% in the two interventions after 2011). Generally, residents' motivation to be involved in interventions is high at first. However, this dwindles over time, and a strategy for maintaining motivation is important. The above results describe a similar pattern. A socio-demographic comparison of respondents who continued to participate in the model district interventions versus those who stopped (after 2011) revealed a significant difference in their respective ages. The average age of the respondents who continued to participate was 47.5, while the average age of those who stopped was 61.7 ($p < 0.05$).

5.3.4 Concluding Remarks

This survey revealed the potential influence of efforts involving residents with support from local governments (e.g., the Satoyama Woodland Improvement Program, the Model District Program) on communities and residents of different districts. I have no access to information about Momurahonden and the neighboring districts prior to the launch of these programs. Accordingly, I can speculate that the programs and interventions helped increase residents' participation and that residents will continue to work on interventions for a certain period (at least 1 year). Participation in interventions is likely to mitigate the perceptions that the damage is serious and improve residents' perceptions of local governments supporting interventions. Similarly, I can infer that such participation encourages residents to take the initiative in wildlife interventions, without relying on local governments. The results showed also the repetition of similar interventions gradually discourages residents' participation in collaborative activities. Senior residents, in particular, are likely to stop participating.

To accurately evaluate programs, basic information (e.g., status of damage and interventions, residents' perceptions, administration by the residents' association) must be collected prior to their launch. In terms of residents' perceptions, a questionnaire survey (similar to the one conducted in this research) should be conducted in advance. Items that are likely to be affected by interventions (e.g., knowledge and skills related to wildlife and interventions, motivation) should be selected and measured ahead of time. In the preliminary phase, it is essential to identify as many comparison groups as possible that have approximate values for both items subject to evaluation (e.g., perceptions of wildlife issues) and other relevant items (e.g., population, number of households, total cultivated area, geographical conditions) (Ryu and Sasaki 2010). In general, the closer to the target district (model district) a district is, the higher the degree of approximation. In this survey, two districts adjacent to the model district were used for comparison. Precautions should also be taken in the evaluation process due to the increased risk of information exchange occurring between the groups (residents of the districts).

In this paper, I evaluated the results of the programs (impact evaluation). To improve the accuracy of this procedure, a theory evaluation (e.g., whether the goals for the programs were set properly in the first place, whether interventions were effective in achieving the goals) and a process evaluation (e.g., whether the initially planned interventions were implemented, whether the expected number of residents participated) could be conducted concurrently (Rossi et al. 2004). In addition, wildlife managers could implement more adaptive measures/interventions and more accurately reflect the changing status of the districts by reflecting the results of the survey (Decker et al. 2012).

References

Agee JD, Miller CA. Factors contributing toward acceptance of lethal control of black bears in Central Georgia, USA. Hum Dimens Wildl. 2009;14:198–205.

Ajzen I. From intentions to actions: a theory of planned behavior. In: Kuhl J, Beckman J, editors. Action control: from cognition to behavior. Heidelberg: Springer; 1985. p. 11–39.

Ajzen I, Fishbein M. Understanding attitudes and predicting social behavior. London: Prentice-Hall International, Inc; 1980. p. 278.

Akiba H, Miller CA, Matsuda H. Factor influencing public preference for raccoon eradication plan in Kanagawa, Japan. Hum Dimens Wildl. 2012;17:207–19.

Baruch-Mordo S, Breck SW, Wilson KR, Broderick J. A tool box half full: how social science can help solve human-wildlife conflict. Hum Dimens Wildl. 2009;14:219–23.

Carpenter LH, Decker DJ, Lipscomb JF. Stakeholder acceptance capacity in wildlife management. Hum Dimens Wildl. 2000;5:5–19.

Decker DJ, Purdy KG. Toward a concept of wildlife acceptance capacity in wildlife management. Wildl Soc Bull. 1988;16:53–7.

Decker DJ, Brown TL, Siemer WF. Human dimensions of wildlife management in North America. Bethesda: The Wildlife Society; 2001, 447 pp

Decker DJ, Riley SJ, Siemer WF. Human dimensions of wildlife management. 2nd ed. Baltimore: The Johns Hopkins University Press; 2012.

Enari H, Suzuki T. Risk of agricultural and property damage associated with the recovery of Japanese monkey populations. Landsc Urban Plan. 2010;97:83–91.

Ericsson G, Heberlein T. Attitudes of hunters, locals, and the general public in Sweden now that the wolves are back. Biol Conserv. 2003;111:149–59.

Ferraro PJ, Pattanayak SK. Money for nothing? A call for empirical evaluation of biodiversity conservation investments. Public Lib Sci Biol. 2006;4(4):482–8.

Gigliotti L, Decker DJ, Carpenter LH. Developing the wildlife stakeholder acceptance capacity concept: research needed. Hum Dimens Wildl. 2000;5:76–82.

Hashimoto K, Ko K, Fujinaga H, Lutz R. Predictive ability of the theory of planned behavior for mental health outcomes in Japanese vs. Chinese students. J Health Sci. 2008;30:27–37. (in Japanese with English abstract)

Hayabuchi Y. Spillover effect of environmental education. Kyoto: Nakanishiya Press; 2008. (in Japanese)

Hirose Y. Determinants of environment-conscious behavior. Res Soc Psychol. 1994;10(1):44–55. (in Japanese with English abstract).

Hirose Y, Kobayashi H, Shima T. Clarification of the impact of determinant of preference of non-farmers to the Irrigation Water Wheels on motive of conservation action. J Rural Plan. 2010;32:287–92. (in Japanese with English abstract)

Honda Y. Factors affecting non-farmers' participation in maintenance of irrigation and drainage canals: focusing on styles of irrigation water management and canal maintenance. J Rural Plan. 2011;30:74–82. (in Japanese with English abstract)

Hrubes D, Ajzen I, Daigle J. Predicting hunting intentions and behavior: an application of the theory of planned behavior. Leis Sci. 2001;23:165–78.
Jonker SA, Organ JF, Muth RM, Zwick RR, Siemer WF. Stakeholder norms toward beaver management in Massachusetts. J Wildl Manag. 2009;73:1158–65.
Kanuma City. Agricultural damage of Kanuma City. 2008. (in Japanese). http://www.city.kanuma.tochigi.jp/resources/content/1943/keikaku.pdf. Accessed 29 Oct 2012.
Kanuma City. Data of Kanuma city. 2010. (in Japanese). http://www.city.kanuma.tochigi.jp/13,20117,147,510.html. Accessed 5 Nov 2013.
Kikuchi N. Revived oriental white stork. Tokyo: Tokyo University; 2006. (in Japanese)
Kimura M, Miyagi A, Ide S. Key factors in sustaining river conservation activities by community associations – case of Jichikais' in Moriyama, Shiga. Environ Syst Res. 2008;36:475–83. (in Japanese with English abstract)
Kinoshita D, Kuki Y, Hoshino S, Takeyama E. Present state of the group activities to decrease agricultural damage caused by wildlife and possibility of non-farmers' cooperation on the activities in paddy areas: cases study in Nantan region, Kyoto Prefecture. J Rural Plan. 2009;27:227–32. (in Japanese with English abstract)
Krebs, C. J. 2001. Ecology: Fifth edition. Benjamin Cummings, San Francisco., , 695 pp.
Majic A, Bodonia AMT, Huber D, Bunnefeld N. Dynamics of public attitudes toward bears and the role of bear hunting in Croatia. Biol Conserv. 2011;144:3018–27.
Martin SR, McCurdy K. Wilderness food storage in Yosemite: using the theory of planned behavior to understand backpacker canister use. Hum Dimens Wildl. 2009;14:206–18.
Maruyama T. Study about macaques conservation and management system at cities, towns, and villages. Yaseichojyu Kenkyu Kiyo (Bull Wildl Tochigi Prefecture). 2005;31:34–40. (in Japanese)
Maruyama Y. Environmental issues of macaques and people: between nature conservation and damage regarding Japanese macaques. Kyoto: Showado; 2006. (in Japanese).
Ministry of Agriculture, Forestry and Fisheries. Agricultural damage by wildlife in each prefecture. 2007. (in Japanese). http://www.maff.go.jp/j/seisan/tyozyu/higai/h_zyokyo/h19/pdf/ref_data02.pdf. Accessed 8 Feb 2018.
Ministry of Agriculture, Forestry and Fisheries. Agricultural damage by wildlife in each prefecture. 2015. (in Japanese). http://www.maff.go.jp/j/seisan/tyozyu/higai/h_zyokyo2/h27/attach/pdf/h27-2.pdf. Accessed 22 Jan 2015.
Misaka K. Development of a model of psychological processes for promotion of environmental education. Environ Educ. 2003;13(1):3–14. (in Japanese with English abstract)
Mizuhara T. Introduction to social psychology. The second edition. Theory and practice. Tokyo: University of Tokyo Press; 1989. 260pp. (in Japanese).
Murase Y, Takada H, Hirose T. Multivariate statistics using SPSS. Tokyo: Ohmsha; 2009. 351pp. (in Japanese)
Nasushiobara City. Statistical record of Nasushiobara city: 2010. 2010. https://www.city.nasushiobara.lg.jp/dbps_data/_material_/_files/000/000/000/930/4dbe5647004.pdf. Accessed 29 Dec 2011.
Nasushiobara City. Statistical record of Nasushiobara city: 2010. 2011. (in Japanese). http://www.city.nasushiobara.lg.jp/191/3094/2934/index.html. Accessed 6 June 2011.
Niibe K, Yano Y, Matsuda N, Takahashi Y, Maruyama T. Population of and damage caused by large mammals based on questionnaire survey. Yaseichojyu Kenkyu Kiyo (Bull Wildl Tochigi Prefecture). 2008;34:39–47. (in Japanese)
Nikko City. Data of the city. 2010. http://www.city.nikko.lg.jp/seisaku/profile/data/. (in Japanese). Accessed 3 Nov 2013.
NTT East Japan. Hello Page: Nasushiobara District of Tochigi Prefecture. Individual names. Tokyo: NTT East Japan; 2011.
Ohtomo S. Dual decision making process of environment-risk behavior and the influence of prototype perception of the environmentally nonconscious person. Environ Educ. 2004;13(2):25–34. (in Japanese with English abstract).

Pouta E, Rekola M. The theory of planned behavior in predicting willingness to pay for abatement of forest regeneration. Soc Nat Resour. 2001;14:93–106.
Riley SJ, Decker DJ. Risk perception as a factor in wildlife stakeholder acceptance capacity for cougars in Montana. Hum Dimens Wildl. 2000;5:50–62.
Roskaft E, Bjerke T, Kaltenborn B, Linnell JDC, Anderson R. Patterns of self-reported fear towards large carnivores among the Norwegian public. Evol Hum Behav. 2003;24:184–98.
Rossi AN, Armstrong JB. Theory of reasoned action vs. theory of planned behavior: testing the suitability and sufficiency of a popular behavior model using hunting intentions. Hum Dimens Wildl. 1999;4:40–56.
Rossi PH, Lipsey MW, Freeman HE. Evaluation: a systematic approach. 7th ed. Thousand Oaks: Sage Publication; 2004.
Ryu Y, Sasaki R. Evaluation theories and techniques. Tokyo: Taga Shuppan; 2010. p. 217. (in Japanese)
Sakurai R. Chapter 9: Collaborative approach for coexistence with wildlife in rural regions of Japan. In: Frank B, Gilkman JA, Marchini S, editors. Human-wildlife interactions: turning conflict into coexistence. Cambridge: Cambridge University Press; 2019.
Sakurai R, Ueda G, Jacobson SK. Survey of residents regarding the Asiatic black bear at Tajima region of Hyogo Prefecture: designing and conducting questionnaire survey for policy application and management interventions. Wildl Conserv Jpn. 2012a;12(2):33–46. (in Japanese with English abstract)
Sakurai R, Ueda G, Jacobson SK. Effectiveness of the community bear education seminar – conducting pre and post survey at Hidaka Town of Toyooka City in Hyogo Prefecture. Kyosei Stud. 2012b;6:380–92. (in Japanese with English abstract)
Sakurai R, Matsuda N, Maruyama T, Jacobson SK. Overview of the model district program for reducing human-wildlife conflicts in Tochigi Prefecture. Wildl Hum Soc. 2013a;1:47–54. (in Japanese with English abstract)
Sakurai R, Jacobson SK, Ueda G. Public perceptions of risk and government performance regarding bear management in Japan. Ursus. 2013b;24:70–82.
Sakurai R, Enari H, Matsuda N, Maruyama T. Testing social-psychological theories to predict residents' behavioral intentions regarding wildlife issues – application of theory of planned behavior and wildlife acceptance capacity model. Honyurui Kagaku (Mamm Sci). 2014a;54(2):219–30. (in Japanese with English abstract)
Sakurai R, Matsuda N, Maruyama T, Takahashi Y. Residents' perceptions of wildlife conflicts at Momurahonden of Nasushiobara city: comparison with previous year and neighbor communities. Yaseichojyu Kenkyu Kiyo (Bull Wildl Tochigi Prefecture). 2014b;39:61–6. (in Japanese)
Sakurai R, Jacobson SK, Ueda G. Public perceptions of significant wildlife in Hyogo, Japan. Hum Dimens Wildl. 2014c;19:88–95.
Sorice MG, Conner JR. Predicting private landowner intentions to enroll in an incentive program to protect endangered species. Hum Dimens Wildl. 2010;15:77–89.
Sudo H. Determinants of exercise adherence of middle-aged women in the gym. Sport Sci Res. 2008;5:96–107. (in Japanese).
Suwa H, Yamamoto H, Okada I, Ohta T. Path analysis model for development of environmental education program to promote environment-conscious behavior. J Soc Inf. 2006;18(1):59–70. (in Japanese with English abstract).
The Nature Conservation Society of Japan. Wildlife conservation and law from ecological perspectives: toward conservation of biological diversity. Tokyo: Kodansha; 2010. (in Japanese)
Tochigi Prefecture. Biodiversity and Tochigi Strategy: connecting and coexisting lives at Tochigi. Tochigi: Nature Preservation Division, Department of Environment and Forestry, Tochigi Prefecture; 2010. (in Japanese)
Utsunomiya University. The Satoyama wildlife management technologist training program. 2012. (in Japanese). http://ssrc.utsunomiya-u.ac.jp/wmtp/. Accessed 23 Mar 2012.

Watanabe O, Ogura S. Relationship between perceptions toward wildlife values and opinions about wildlife management policy in rural areas of central Japan. Wildl Conserv Jpn. 1996;2(1):1–15. (in Japanese with English abstract)

Yamabata N. Mitigate effect on damage to food crops achieved by collaboration of a whole village for chase-off of monkeys: verification in 6 area of Mie Prefecture. J Rural Plan. 2009;28:273–8. (in Japanese with English abstract)

Yamabata N. Effect of improved countermeasures to agricultural damage by wildlife on farmers' awareness of farmland management: a study of community in Mie Prefecture. J Rural Plan. 2010;29:245–60. (in Japanese with English abstract).

Yamabata N. Effect of chasing away by village on the home range and appearances of a macaques group: verification in 7 area of Mie Prefecture. J Rural Plan. 2011;30:381–6. (in Japanese with English abstract).

Yamamoto T, Nagasawa T. Motivation factors of local residents' participation in the maintenance works of irrigation and drainage facilities. J Rural Plan. 2010;28:387–92. (in Japanese with English abstract).

Yorifuji K, Hirose Y. Determinants of children's waste-reduction behavior. Environ Educ. 2002;12(1):26–36. (in Japanese with English abstract)

Zinn HC, Manfredo MJ, Vaske JJ. Social psychological bases for stakeholder acceptance capacity. Hum Dimens Wildl. 2000;5:20–33.

Chapter 6
Introducing Ideas and Approaches of Human Dimensions in Japan

Abstract This chapter explains how conducting human dimensions studies has contributed to improved wildlife management in Japan. The chapter begins with an explanation of the meaning and significance of human dimensions studies, e.g., how anecdotal assumptions shared among stakeholders could be tested scientifically using a human dimensions approach. There follows an examination of useful tools and ideas for exposing wildlife issues and displaying research results (wildlife acceptance capacity, social psychological theories, the Potential for Conflict Index, and the logic model). The second part of the chapter discusses aspects to be considered when conducting human dimensions studies in specific cultural and social settings. Section 6.2.1 explains the significance of conducting research in collaboration with local stakeholders and governments. The final part of the chapter considers the respective strengths of qualitative and quantitative studies.

6.1 How Can Human Dimensions Contribute to Solving Human-Wildlife Conflicts?

6.1.1 The Significance of Human Dimensions: Changing Empirical Knowledge to Research-Based Facts

An important element of human dimensions research is its ability to unveil new perspectives related to the social aspects of wildlife management. While human dimensions studies in Japan have traditionally been scarce, as explained in Sect. 3.4, both researchers (conventionally those working in fields such as ecology and veterinary studies) and government officials recognize the potential impacts of, for instance, the media in forming public perceptions or influencing public attitudes with regard to management or decision-making. In other words, despite the lack of official research, empirical knowledge is shared among researchers and government officials regarding the necessity of understanding human dimensions factors to improve wildlife management. As explained in Chaps. 4 and 5, introducing the ideas and concepts behind human dimensions, as well as actually conducting human

R. Sakurai, *Human Dimensions of Wildlife Management in Japan*, Ecological Research Monographs, https://doi.org/10.1007/978-981-13-6332-0_6

dimensions studies in Japan, has provided opportunities to other researchers and government officials to examine social aspects that may become the target of future research. Existing human dimensions studies have demonstrated that those shared anecdotal assumptions can be tested scientifically.

Historically, only a limited number of human dimensions studies have been conducted in Japan. There are several reasons for this. For instance, the methods and tools used in human dimensions studies are not shared widely among Japanese stakeholders because of the limited number of universities or laboratories where students can learn about human dimensions in relation to wildlife. There are universities in Japan where professors affiliated to the college of social science or the faculty of sociology teach sociological methods and theories that could be applied to wildlife management. If students in the wildlife management faculties or departments adopted these methods in their research or if students in the faculties of social science or sociology decided to expand their research to wildlife management, the amount of human dimensions research could be gradually increased. Moreover, if students (or researchers and government officials) studied the discipline as it has been developed in the USA (such as by visiting and studying at American universities), this could increase the number of human dimensions researchers in Japan.

The media analysis in Sect. 4.2 revealed that a majority of newspaper articles regarding black bears referred to people's anxieties about an increase in bear conflicts. The number of articles that mentioned the negative aspects of human-bear interactions far outweighed those that referenced positive aspects linked to the bears' presence. Media analysis is recognized as an applicable scientific approach in the field of wildlife management in Japan, with results presented at academic conferences. Other researchers might be encouraged to apply a similar approach to their own studies. After I presented my media analysis of wildlife to the Wildlife Conservation Society (currently the Association of Wildlife and Human Society) in 2011 (Sakurai 2011), similar conference presentations have been made on an almost yearly basis (e.g., Okuda et al. 2013; Mitsui et al. 2016; Tashiro 2016).

Human dimensions studies, such as research into public perceptions, have been frequently conducted and presented in Japan by researchers and governmental officials in various wildlife-related conferences, including the Mammal Society of Japan and the Society for Environmental Economics and Policy Studies. Similarly, papers introducing the concept of human dimensions in US wildlife or those explaining the results of human dimensions research have been published in a range of Japanese academic journals (e.g., Kubo and Shoji 2012, 2014; Ueda et al. 2012, 2013). As the field of human dimensions broadens and more data are accumulated, it seems increasingly likely that the discipline will achieve recognition in Japan as a valid academic approach. Increasingly, many topics and assumptions that were formerly shared anecdotally among stakeholders (e.g., people's perceptions about wildlife and how the media reports human-wildlife interactions) are discussed as scientific findings derived from quantitative and qualitative analysis.

6.1.2 Some Useful Tools and Ideas

6.1.2.1 Wildlife Acceptance Capacity and Social Psychological Theory

Wildlife acceptance capacity is a new concept in Japan that has yet to be scientifically discussed or evaluated. The rapid increase of certain species, such as deer and wild boars, as well as the high levels of agricultural damage caused by wildlife, has led local residents to request that the numbers of these species be reduced (or in some cases, completely eradicated within the region). Governments encourage local residents to take precautions to prevent agricultural damage, such as erecting electric fences. In many regions, hunters are given incentives to capture overpopulated species according to a Specified Wildlife Management Plan. Another approach toward mitigating human-wildlife conflicts, however, is to increase residents' tolerance toward wildlife. The wildlife acceptance capacity concept implies that there are other policy goals that governments, local residents, researchers, and stakeholders can strive for. Such objectives are aimed not only at controlling the wildlife population and preventing damage but also at increasing people's acceptance capacity, with a view to arriving at an acceptable level of human-wildlife interactions (Decker et al. 2001, p. 125).

Section 5.2 demonstrated how people's risk perceptions and attitudes toward wildlife affect their acceptance capacity. These factors could hence be targets for awareness-raising/education campaigns aimed at increasing tolerance and positive attitudes toward wildlife.

In Sect. 5.2, carrying on from where previous studies (Riley and Decker 2000) left off, wildlife acceptance capacity was measured on a three-point scale whereby residents were asked whether the population of wild boars should be decreased, stabilized, or increased. However, I consider this scale to be overly simplistic. Wildlife acceptance capacity could be more accurately measured by asking various related questions. At this instance, additional questions might be related to the frequencies with which local residents encountered wild boars (e.g., in the past 2 years), the place where these encounters occurred (near a residential area, etc.), or how often they were prepared to tolerate encounters with wild boar (e.g., how many times per month, per year). In addition, it would be important to monitor how wildlife acceptance capacity increased in the aftermath of awareness-raising/education activities. However, research into the short- and long-term changes in public wildlife acceptance capacity, as well as the effect of outreach programs on tolerance, is notably lacking (Sakurai 2019). These questions remain legitimate topics for future studies.

Social psychological theory, in particular the Theory of Planned Behavior discussed in Sect. 5.2, considers the unique conceptual structure of local residents. Significantly, the study in this chapter was designed and carried out according to theoretical foundations, which enabled the specific findings to be extended to a general context (Decker et al. 2001). When I first began human dimensions studies in Japan, it was not uncommon to hear comments from sociologists (particularly those preferring a qualitative approach) such as, "since Japan has a totally different cultural and social background, theories developed in North America cannot explain the current situation. It is therefore meaningless to test these theories." I also heard several ecologists say that human-wildlife conflicts occur simply because local residents lack the necessary knowledge. However, as explained in Sect. 5.2, independent factors comprising the Theory of Planned Behavior may explain residents' behavioral intentions even in Japan. The model, including additional factors, successfully explained nearly half of all related variables. This implies that the behaviors and thought processes of people across different cultures and societies are based on a similar conceptual mechanism. Thus, research results and suggestions from other countries are certainly applicable in Japan. The survey also revealed the unique characteristics of the study sites and that social norms had by far the strongest impact on local residents' behavioral intentions (i.e., people's disinclination to engage in wildlife damage prevention interventions was not necessary because they were ignorant of the issues or they lacked knowledge but rather because they did not feel any social pressure to do so). These are important findings that should be utilized when designing effective awareness-raising/education activities to encourage local residents' involvement in mitigating wildlife conflicts.

Social trust is another social psychological concept that has been tested and utilized in human dimensions research in North America (Vaske et al. 2004; Needham and Vaske 2008). However, it has yet to be applied to wildlife management in Japan. In Sects. 4.3 and 5.3, social trust was measured by evaluating people's perceptions of government performance. The results provided new insights on this issue, revealing that people's level of trust toward government may affect their behavioral intentions. It is, hence, important for governments to build trust relationships with local residents. Listening to residents' concerns and demands and communicating directly with them are ways in which this could be achieved. A series of seminars conducted by local government (see Sect. 4.5) potentially increased residents' trust of the authorities and consequentially their damage prevention behaviors. Similarly, the improvements in trust and preventative behavior arising from the participatory collaborations between residents and government described in Sect. 5.3 indicate that the authorities are on the right track in terms of mitigating overall human-wildlife conflicts.

6.1.2.2 Potential for Conflict Index

One of the tools developed in the field of human dimensions of wildlife management in North America is the Potential for Conflict Index (PCI). PCI examines and graphically displays the perceptions or acceptance capacity of wildlife, helping managers and stakeholders to interpret statistical data. It illustrates the central tendencies and variations in responses by providing scores on a scale between 0 (when all respondents answered in the same way) and 1 (when the responses were equally divided among two extreme answers). Using PCI, wildlife managers can understand and predict situations in which there may be conflicts among stakeholders (i.e., when the PCI score is near 1). (For more details about PCI, see Manfredo et al. 2003; Vaske 2008; Vaske et al. 2010).

A second-generation version of PCI was used here to understand local residents' tolerance toward six different wildlife species in Japan, with data taken from a survey conducted in Tato Town in Toyooka City (Hyogo Prefecture) in 2010. The species considered were wild boars, deer, macaques (a native species that frequently cause agricultural damage), black bears, the reintroduced Oriental white stork (an endangered species), and nutria (*Myocastor coypus*) (a non-native, invasive species). For each species, residents were asked whether they "should be extirpated" (score = 0), "can exist if they stay in mountains" (score = 1), "can live nearby if they do not cause damage" (score = 2), and "should exist everywhere" (score = 3). These scores were used as dependent variables. Questionnaires were sent to 1,793 households, and 868 were returned (response rate = 48%). [For more details about this study, see Sakurai et al. 2014.]

Figure 6.1 shows the PCI displaying the average scores and variances for residents' perceptions of the six species. The figure illustrates that average scores for wild boars, deer, macaques, and black bears were similar (near "can exist if they stay in mountains"). However, the average score for nutria was located between 0 ("should be extirpated") and 1 ("can exist if they stay in mountains"). The negative score here implies that many residents consider this species to be invasive and that it should be eradicated.

The PCI scores (indicated by the size of the bubbles) were higher for deer and oriental stork. As deer cause more agricultural damage than do other species in the study sites, residents' perceptions ranged from the extreme "should be extirpated" to "can live nearby if they do not cause damage," resulting in a higher PCI score. While many residents felt that the Oriental stork "should live everywhere" or "can live nearby if they do not cause damage," the high PCI score indicates that perceptions were not homogeneous. The divided opinions for the Oriental stork imply that

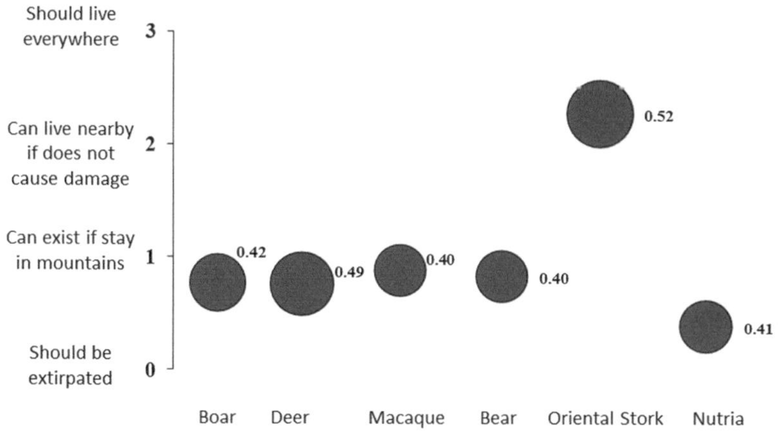

Fig. 6.1 Potential for Conflict Index (PCI). The bubbles show the average scores and variances for residents' perceptions of the six species. (From Sakurai et al. 2014)

all residents may not agree with current conservation policy regarding their reintroduction. Previous studies have found that some residents felt that the city government was spending too much money on efforts to reintroduce the Oriental stork (Honda 2006) at the expense of other issues such as the lack of public transportation in the region.

This was one of the first efforts in Japan to measure PCI in the field of wildlife management. It revealed that people's perceptions of different species can be compared and visually displayed using the same criteria. When presenting PCI to government officials, I felt that it helped them to understand public perceptions in a simple and effective way that avoided the need for statistical jargon. However, for PCI to be of use in the decision-making process, questions would have to be asked regarding more detailed items, such as residents' opinions about a specific policy (e.g., How should a bear captured near a residential area be dealt with? (1) Release the bear after negative conditioning; (2) release the bear without negative conditioning; and (3) destroy the bear). Measuring PCI for specific policies would allow government officials to understand which policies and interventions are preferred by residents. Authorities would also be able to foresee situations that are likely to lead to disagreement and opposition by residents or conflict among stakeholders.

6.1.2.3 Logic Model for Program Evaluation and Improvement

Various programs (e.g., activities and interventions) have been implemented in local communities and by governments to mitigate human-wildlife conflicts. To ensure that such programs and policies are achieving their goals, however, it is essential to display and monitor their pathways/road maps. The logic model is a tool that systematically organizes the resources required for a program, detailing all interventions, expected changes, and results. The model visually displays the program's objectives and expected outcomes. It is hence a useful tool with which to communicate the program's central focus to stakeholders, including organizers, beneficiaries, and evaluation personnel. The logic model is one of the tools most frequently used when designing awareness-raising/education programs and activities in North America (Ernst et al. 2009; Jacobson 2009; Phillips et al. 2012). However, it is rarely employed in Japan, particularly in the field of wildlife management.

I carried out a preliminary survey of local residents before the commencement of the planned Model Program in Ogaki district in Tochigi City, one of the model districts of Tochigi Prefecture. A logic model was then created on the basis of these results. The goal of the logic model in this study was to present a vision for effectively implementing and evaluating the program. Ogaki district was designated a model district in FY2011, and I conducted the survey in the same year, prior to the start of the program. Questionnaires were delivered to all households (n = 248), with 177 responses (response rate = 71.4%).

Table 6.1 presents the logic model that was created on the basis of the results of this survey. [The goal of this chapter is to introduce the logic model. For more information about the detailed results of the survey, see Sakurai et al. 2012.] The logic

Table 6.1 A logic model created on the basis of the results of the preliminary questionnaire survey in Ogaki district, Tochigi City (created based on Sakurai et al. 2012)

Investment	Interventions (implemented every year)	Target number of participants	Short-term goal (one year after commencement of interventions): changes in perceptions	Mid-term goal (three years after commencement of interventions): changes in behaviors	Long-term goal (10 years after commencement of interventions): changes in the community
Stakeholders: • Nature Preservation Division of the Tochigi Prefectural Government • Utsunomiya University • City of Tochigi • Rural Areas Promotion Division of the Tochigi Prefectural Government *Budget:* • Nature Preservation Division (fees to dispatch instructors)	• Workshops about interventions to prevent damage by wild boars • Community inspections within the district • Weeding around agricultural land and residential areas • Study meeting to reduce human–wildlife conflicts	40 40 40 40	• Residents who are willing to work on damage prevention interventions against wild boar increase from 49% to 70% • Residents who know how to prevent damage by wild boars increase from 12% to 50% • Residents who consider that the prefectural government has properly explained how to cope with wild boar issues increase from 10% to 40% • Residents who answer to test questions about damage prevention interventions against wild boars correctly increase from 40% to 70%	• Residents who work on wildlife interventions increase from 61% to 90% • Residents in the Ogaki district outperform other communities in terms of participation in interventions, awareness, and knowledge	• Residents work on damage control interventions on their own initiative and create a community environment with minimal wildlife encounters • Wildlife damage decrease

Status:
Increased encounters with wildlife and wildlife damage in and around the district. Aging and depopulation of the district.
Hypothesis:
Participatory education activities help increase residents' awareness and knowledge and motivate them to work on interventions. Residents' active participation in interventions helps create a community environment with minimal wildlife encounters and damage.

model was created on the basis of the assumption that interventions would be implemented four times a year through the Model Program and that residents' awareness about damage prevention interventions would increase. The aim was that the local community would be able to successfully mitigate wildlife conflicts. One of the stated short-term goals, for instance, was that "residents who are willing to work on damage prevention interventions against wild boars increase from 49% to 70%." The preliminary survey revealed that 49% of respondents were willing to engage in interventions. It was, thus, hypothesized that this figure could be increased to 70% over the course of the year through various participatory and education activities. Long-term monitoring of residents' awareness and knowledge, as well as their actual damage prevention behaviors, was required to verify the results.

The aim was that the logic model would provide both government officials and local residents with a clear understanding of future interventions and the program's goal. To evaluate the impacts of the program and to understand its progress, it is essential to set specific, measurable targets. The logic model shown in Table 6.1 serves as one of these indices. However, the possibility remains that certain interventions may not be implemented on the ground or residents might not participate as planned. At such instances, the model could be revised to reflect the real-life situation and to meet stakeholders' needs where necessary. The logic model is a communicative tool by which to share survey results. Its broader function is to mitigate human-wildlife conflicts and assist program organizers and participants in meeting the program's overall targets.

6.2 Aspects to Be Considered When Conducting Human Dimensions Studies in Specific Cultural and Social Settings

6.2.1 Significance of Collaborative Research with Local Stakeholders and Government

When designing and planning a human dimensions study for human-wildlife conflicts, researchers could look to previous studies or government reports to understand the situation and background of the study sites. In many cases, however, extant research is limited (e.g., local residents' perceptions of specific wildlife or their intentions to engage in wildlife damage prevention interventions), particularly in countries such as Japan where these types of studies remain infrequent. My plan was to study local residents' perceptions of bears in the western part of Japan. However, existing research into this and related topics was scarce. All previous studies were conducted in the eastern part of the country, and I was not sure how the results would correspond to the present study site or whether they were generalizable to other sites at all given that results and conclusions were based on interviews with as few as a dozen residents, with no real explanations as to how those interviewees were selected.

The specific feature of human dimensions research in wildlife management that differentiates it from other similar fields is that it seeks findings that are both scientifically (i.e., academically) and socially (i.e., addressing real-world issues) valuable. As Manfredo and Vaske (1996) wrote in their opening to the first issue of *Human Dimensions of Wildlife*, "human dimensions emphasize the concepts, tools and techniques that give assistance in representing the public and give leadership on wildlife issues." One of the goals is to obtain findings that provide assistance to managers, local residents, and other stakeholders. Any study, the design of which is based solely on previous studies and academic theories, without considering the situation on the ground (e.g., the needs of the local community at the study site), will find it hard to gain knowledge that is useful for stakeholders. In seeking to design meaningful research, researchers should not assume that they know the decision-makers' needs and demands; rather, they should ask questions and discuss issues with these decision-makers and other stakeholders (Decker et al. 2001). Collaboration with local stakeholders and government is hence essential to the design and performance of human dimensions studies.

I first started conducting human dimensions research in Nagano Prefecture, Central Japan, in 2008. Local residents, local nonprofit organizations, local researchers, and local government entities all kindly provided useful information about wildlife conflicts in the region and people's perceptions and behaviors with regard to wildlife issues. This information was extremely useful when it came to planning my research and drafting research questions that related particularly to the local situation. This experience taught me that as far as possible, any human dimensions studies aimed at solving wildlife conflicts should be designed and implemented in collaboration with local governments and/or local stakeholders. Therefore, most of the studies detailed in Chaps. 4 and 5 take the form of collaborative research with prefectural governments.

It should be noted, however, that the term "collaboration" can mean different things to different researchers. Similarly, the level and degree of this collaboration may vary significantly. Numerous studies have been conducted and coauthored by researchers and governments under the guise of collaborative research. However, my belief is that to implement collaborative research that is both scientifically and socially meaningful, this collaboration must function at every step of the process, including developing research questions and hypotheses, deciding upon study sites and research methods, and polishing the wording of questionnaire items. A rigorous approach such as this may be more time consuming than one that neglects discussion with local stakeholders. Nevertheless, these interactions, when implemented, enhance the research process and have the long-term benefit of providing governments and stakeholders with information that will be of genuine assistance to future decision-making.

For example, in my human dimensions studies of black bear management in Hyogo, prior discussions with governmental officials gave me an understanding of some of the region's urgent issues (people not engaging in interventions), potential behaviors that residents may engage in (damage prevention interventions for their persimmon and chestnut trees), and information sought by government officials (e.g., measurable quantitative outcomes, such as the reporting ratios for each district). The study coincided with a massive number of bears appearing near residential areas across the study sites – the year was 2010. Local government understood the timely nature of the study and realized that the results would be applicable to future decision-making. Similarly, in my capacity as researcher, I was able to utilize tools and concepts such as the Theory of Planned Behavior and wildlife acceptance capacity that would help officials understand and predict residents' perceptions and behaviors.

6.2.2 The Power of Qualitative Study

My research into black bear management in Hyogo (Chap. 4) and wildlife management in Tochigi (Chap. 5) began with a qualitative approach (interviews with stakeholders, participant observation, etc.). Researchers are generally equipped with the knowledge of academic discipline. They make hypotheses on the basis of previous studies and design their research using theories and models that may be conceived before visiting the study site. However, studies designed solely from an academic perspective, which do not consider the situation on the ground, may result in data that is neither applicable nor useful to government officials' decision-making processes. Conversely, qualitative studies that seek out stakeholders' opinions, examine the damage sites, and observe the actual wildlife damage prevention interventions give researchers a nuanced, fruitful understanding of both the people affected and the situation on the ground (Decker et al. 2001). Clearly, and by extension, it is best to talk to as many stakeholders as possible. An example of this would be conducting interviews not only with district leaders or specific stakeholders but also with normal residents. Likewise, it is preferable to talk to government officials from a range of different departments (e.g., both the Nature Preservation Division and the Department of Agricultural Administration).

As explained in Chap. 4, one of the benefits of carrying out qualitative studies before quantitative ones is that in many cases they reveal a bias on the part of the researcher. By talking to stakeholders, certain questionnaire items, for instance, are shown to be inappropriate to the local context. Prior to carrying out my interview about black bears, for example, I did not imagine that individuals could hold such conflicted sentiments about the animals (e.g., feelings of fear and guilt while simultaneously perceiving the bears as "cute").

The advantage of qualitative studies is that researchers can use real-life narratives as data to vividly demonstrate a particular situation or phenomenon. As part of the research described in Sect. 4.3, for instance, I was given a graphic example explaining the relationship between local residents and wildlife in the region: "In Tanto town, a bear entered the town hall building. The staff members were evacuated to the second floor. I was told that a boar went into a coffee house in the town yesterday. People in our community joke that the bear came to the town hall to pick up a residence certificate, and the boar came to the coffee house to have a cup of coffee."

Another strength of the qualitative approach is that by conducting interviews and actively participating in activities, frequent, long-term interactions/collaborations can be established, allowing researchers to build trust with local residents, government officials, and other stakeholders. For the research described in Chap. 4, I rented a house in a mountain village at one of the study sites and lived as a resident of the district. Although it was a temporary setup, I nevertheless joined the residents' association and participated in its regular meetings. My daily interactions with the local residents made them more open to me as time progressed and thus more likely to reveal their honest opinions. As a researcher, my experience of living in the district taught me that wildlife issues were not necessarily the most serious problem confronting the region. Rather, it was one of many problems that the region sought to address, including depopulation, aging, the lack of public transportation, and the lack of hospital facilities.

By conducting interviews with government officials from the town hall, city government, and prefectural government and discussing my research plans with them, they started to realize that the aim of my research was not only to gain scientifically significant data but also to obtain information that would assist with real-life management and decision-making. In the course of these discussion and meetings, I was able to gain the trust of governmental officials, who, in turn, provided me with suggestions and requests to revise my questionnaire survey and even proposed some ideas for potential future research. This collaborative process allowed me to revise and finalize the items of the questionnaire, such that its contents were both scientifically meaningful and culturally/socially appropriate.

6.2.3 *The Power of Quantitative Study*

As indicated above, qualitative studies reveal important aspects of local residents' beliefs and opinions related to wildlife, as well as how their lifestyles are connected to wildlife issues. Quantitative studies, meanwhile, have considerable power in that they are able to explain the local situation using minimal, simple, and accessible information. While the results of qualitative studies have academic significance in providing a narrative description of a specific situation, a simple fact such as "only 31% of residents report the sighting of a bear" may be sufficient to encourage governments to implement new interventions. One of the government's responsibilities is to meet and reflect public needs and demands. Understanding what the majority of residents are thinking and doing is thus indispensable with respect to developing meaningful, democratic policies. Perhaps for this reason, a majority of papers published in *Human Dimensions of Wildlife* from 1996 to 2006 were from a quantitative rather than qualitative perspective (Vaske et al. 2006).

In Sect. 4.4, for example, the results of the Hyogo Prefecture survey regarding bear management helped government officials to understand both the situation on the ground and the importance of certain issues. These results also enabled managers to set practical and realistic goals for future interventions, such as increasing the reporting rate from 40% to 80% over 2 years. Likewise, local governments were inspired to design and implement broad-scale education seminars, for the first time in the prefecture's history. Additional studies evaluating these seminars (Sect. 4.5) revealed that these interventions had a positive impact on participants' damage prevention behaviors. Indeed, one government official from the prefecture mentioned to me in conversation that the educational seminars and other collaborative interventions had brought about a decrease in the number of human-bear conflicts.

6.2.4 Conclusion

These series of human dimensions studies have demonstrated how human dimensions research could potentially help to mitigate or even resolve human-wildlife conflicts in the real world. Understanding the local situation helps managers to design and implement new interventions/policies. Evaluating the impacts of these interventions/policies in turn allows us to understand the effectiveness of the programs and to identify areas for improvement. Used in tandem, qualitative and quantitative methods allow governments to realize the importance and seriousness of the situation. The described mixed-mode results were sufficiently powerful and comprehensible for officials – to the extent that they used their data to implement new policies.

References

Decker DJ, Brown TL, Siemer WF. Human dimensions of wildlife management in North America. Maryland: The Wildlife Society; 2001.

Ernst JA, Monroe MC, Simmons B. Evaluating your environmental education programs: a workbook for practitioners. Washington, D. C: North American Association for Environmental Education; 2009.

Honda Y. Significance in the release of Oriental White Storks (*Ciconia boyciana*) from the viewpoint of local people soon after the release: the questionnaire survey in whole new Toyooka City. Bull Tokyo Univ For. 2006;116:113–143. (in Japanese with English abstract).

Jacobson SK. Communication skills for conservation professional. 2nd ed. Washington, D. C: Island Press; 2009.

Kubo T, Shoji Y. Using choice experiment to assess potential demand for a bear-watching tour; a case study in Daisetsuzan national park, Japan. Wildl Conserv Jpn. 2012;13(2):9–18. (in Japanese with English abstract).

Kubo T, Shoji Y. The role of environmental economics in wildlife management. Rev Environ Econ Policy Stud. 2014;7(1):66–68. (in Japanese).

Manfredo MJ, Vaske JJ. Human dimensions of wildlife: introduction. Hum Dimens Wildl. 1996;1:5–6.

Manfredo MJ, Vaske JJ, Teel TL. The Potential for Conflict Index: a graphic approach to practical significance of human dimensions research. Hum Dimens Wildl. 2003;8:219–28.

Mitsui S, Kubo T, Yoshida M. Content analysis of information on cat management in Ogasawara Islands. Program and abstracts of 22nd conference of Association of Wildlife and Human Society. 2016. p. 132. (in Japanese).

Needham MD, Vaske JJ. Hunter perceptions of similarity and trust in wildlife agencies and personal risk associated with chronic wasting disease. Society and Natural Resources. 2008;21:197–214.

Okuda K, Sakurai R, Okuda K, Enari H, Kodera Y. Newspaper coverage of wildlife management from data of the Japan Agricultural News. Program and abstracts of 19th conference of Association of Wildlife and Human Society. 2013. p. 71–72. (in Japanese).

Phillips T, Bonney R, Shirk JL. What is our impact? Toward a unified framework for evaluating outcomes of citizen science participation. In: Dickinson J, Bonney R, editors. Citizen Science: public participation in environmental research. New York: Cornell University Press; 2012. p. 82–95.

Riley SJ, Decker DJ. Risk perception as a factor in wildlife stakeholder acceptance capacity for cougars in Montana. Hum Dimens Wildl. 2000;5:50–62.

Sakurai R. Evaluating the newspaper coverage regarding black bear issues. Program and abstracts of 17th conference of Wildlife Conservation Society. 2011. p. 103–104. (in Japanese).

Sakurai R. Collaborative approach for coexistence with wildlife in rural regions of Japan. In: Frank B, Gilkman JA, Marchini S, editors. Human-wildlife interactions: turning conflict into coexistence. Cambridge: Cambridge University Press; 2019.

Sakurai R, Matsuda N, Maruyama T, Takahashi Y. Front-end evaluation of the Model District program for reducing human-wildlife conflicts at Ogaki district of Tochigi city: preparation and application of a logic model. Wildl Res Bull. 2012;38:22–28. (in Japanese).

Sakurai R, Jacobson SK, Ueda G. Public perceptions of significant wildlife in Hyogo, Japan. Hum Dimens Wildl. 2014;19:88–95.

Tashiro Y. History of conservation of endangered species *Kawabatamoroko* using newspaper articles. Program and abstracts of 22th conference of Association of Wildlife and Human Society. 2016. p. 129. (in Japanese).

Ueda G, Kodera Y, Kurumada T, Takeuchi M, Sakurai R, Sasaki C. Factors affecting hunting desertion in Japan: proposal for retention measure. Wildl Conserv Jpn. 2012;13(2):47–57. (in Japanese with English abstract).

Ueda G, Abe G, Sakata H. Evaluation of a program to improve the capture efficiency of feeding traps. Mamm Sci. 2013;53(1):31–42. (in Japanese with English abstract).

Vaske JJ. Survey research and analysis: applications in parks, recreation and human dimensions. Pennsylvania: Venture; 2008.

Vaske JJ, Timmons NR, Beaman J, Petchenik J. Chronic wasting disease in Wisconsin: Hunter behaviors, perceived risk, and agency trust. Hum Dimens Wildl. 2004;9:193–209.

Vaske JJ, Shelby LB, Manfredo MJ. Bibliometric reflections on the first decade of human dimensions of wildlife. Hum Dimens Wildl. 2006;11:79–87.

Vaske JJ, Beaman J, Barreto H, Manfredo MJ. An extension and further validation of the Potential for Conflict Index. Leis Sci. 2010;32:240–54.

Chapter 7
Educating International Students About the Human Dimensions of Wildlife Management

Abstract Wildlife managers should be able to understand stakeholders' needs, motivations, and interests to successfully perform their role. Educating future wildlife managers/researchers about the necessary concepts, tools, and ideas relating to human dimensions is an important element in the achievement of this goal. This chapter explains the key concepts and aims of human dimensions education. A case study of a human dimensions university course taught by the author entitled "Human Dimensions of Nature Conservation" is described and discussed. Feedback indicated that students learned about human wildlife issues, as well as ways to solve problems relating to human-wildlife conflicts. They also acquired important communication skills via mock interview activities and workshops. Finally, the significance of human dimensions education is discussed, as well as potential ways that the course might be improved.

7.1 Background

Human-wildlife issues occur all over the world. Educating future wildlife managers/researchers about the concepts, tools, and ideas related to both ecological aspects and human dimensions is hence increasingly important. According to Decker et al. (2012), human dimensions education has three fundamental goals: (1) to help future professionals understand the factors that influence people's values, beliefs, attitudes, decisions, and actions; (2) to help future professionals understand organizations, communities, and institutions; and (3) to build skills that professionals need to work effectively with co-workers, partners, or stakeholders. Modern wildlife management frequently involves teams whose members seek to find solutions to important issues for sustainable community-based management. Wildlife managers hence need to have the capacity to facilitate or lead the participatory process (Decker et al. 2001). To ensure the success of all stages of wildlife management, managers need to understand stakeholders' (e.g., local residents) needs, motivations, and interests and develop a sense of mutual trust. Future wildlife managers and researchers must recognize the need to incorporate human components into all stages of wildlife management. Similarly, they must understand that lifelong human

R. Sakurai, *Human Dimensions of Wildlife Management in Japan*, Ecological Research Monographs, https://doi.org/10.1007/978-981-13-6332-0_7

dimensions education is indispensable to encouraging and facilitating effective management practices (Jacobson and McDuff 1998; Decker et al. 2012). Decision-making, stakeholder involvement, policy processes, program evaluation, planning, social assessment, normative ethics, law and policies, and communication are all important areas that human dimensions education should seek to incorporate.

7.2 Designing a Human Dimensions of Wildlife Management Course for Japanese Universities

I have been fortunate to teach a human dimensions of wildlife management course to international students from all over the world at two universities in Japan (Chiba University and Ritsumeikan University).

This course is currently offered to undergraduate students at the Ritsumeikan University College of Policy Science. Students are expected to acquire academic knowledge in several disciplines, with a focus on solving real-life problems (related to any social issues) via the acquisition of policy design and implementation skills. The Community and Regional Policy Study Major (CRPS) is taught in English in the College of Policy Science. Students accruing more than 124 credits over 4 years graduate with a Bachelor of Policy Science degree.

As of July 2018, 86 students from all over the world are enrolled in the CRPS (26 freshmen, 26 sophomores, 23 juniors, and 11 seniors). Most come from East Asia (e.g., China, Korea, India, and Indonesia), with a few from Europe and North America. The course I teach (human dimensions of nature conservation) is offered as one of the university's Vision Building Subjects. These courses address a wide range of policy issues, and all students from freshmen to seniors are eligible to register. The goal of the course is to educate students on (1) human-wildlife conflicts both in Japan and worldwide, including causes and background issues, and (2) key factors and ideas for understanding and solving human-wildlife conflicts, including tools and the human dimensions approach.

To offer maximum benefit to students, the course was designed to include both lectures and active learning activities, including group discussions, interviews, debates, guest lectures, and workshops.

When addressing the topic of human dimensions, it is important to first understand the basics of the relationship between human society and natural resources. In the first 3 weeks, the course covers topics including the values of natural resources, types of ecosystem services, biodiversity, threats to biodiversity, and endangered species (see Table 7.1 below).

Weeks 4–10 cover topics including wildlife species in Japan and issues facing rural areas (depopulation, aging, damage caused by wildlife, etc.), case studies of efforts to mitigate wildlife issues (students' involvement, hunting etc.), and case studies from around the world (efforts to utilize wildlife for community development, etc.).

Table 7.1 Overall schedule and contents of the human dimensions of nature conservation course offered at the Ritsumeikan University College of Policy Science

Week	Contents
1	Introduction; what is nature?
2	Human life and natural environment; value of natural resources, biodiversity, ecosystem services
	Conservation of nature; history of environmental issues in Japan, framework for natural resource management
3	Conservation of endangered species
4	Wildlife in Japan; number of species in Japan, the population trends of bears, wild boars, deer, and macaques
5	Rural villages and wildlife conflicts; depopulation and aging in local communities, human-bear conflicts
6	Students' involvement in wildlife issues; student rangers for mitigating wildlife damage
7	Hunting; rapid decrease and aging of hunters in Japan
	Explanation of the debate (preparing for the debate)
8	Group work about the debate
9	Debate
10	Living with wildlife; cases from around the world
11	Academic discipline of Human Dimensions of Wildlife Management Survey and interview to encourage collaboration; preparing for the interview exercise
12	Interview exercise
	Lecture about facilitation and workshop; importance of reaching consensus
13	Preparation for the workshop; group work
14	Facilitation and workshop; solving human-bear conflicts
15	Workshop review; presentation by evaluation group

Weeks 11–15 cover concepts and ideas related to the academic discipline of human dimensions of wildlife management, including social scientific tools for understanding stakeholders' perceptions (interviews and surveys) and tools for building stakeholder consensus (workshops and facilitation).

To increase the educational scope, I included active learning activities in addition to the regular lectures, allowing students to learn from one another. In week 9, for instance, students were divided into two groups and asked to take opposing stances to a certain topic, which they would later debate. Students chose one topic from either A ("The reintroduction of wolves into Japan") or B ("The eradication or raccoons (invasive species) from Japan"). Students had to state whether they agreed or disagreed and take a side. They were then given 2 weeks to study the topic and compile evidence to support their arguments. For the debate itself, each side was given equal time to set out their arguments using data and literature from their research. At the end, the audience (other students) and the instructor decided on the winner.

In week 12, students completed interviewing exercises. Each student was asked to interview other students on a certain topic, e.g., whether they felt there was sufficient greenery on campus and what changes, if any, should be implemented.

Fig. 7.1 Role-playing workshop hosted by students: a core activity of the human dimensions of nature conservation course at Ritsumeikan University (2017)

Students were given a week to prepare their questions. After the interviews were completed, they compiled and summarized the data, which they used to draft suggestions based on their findings.

Also in week 12, after learning about the methods and goals behind facilitation measures and workshops, students hosted a workshop in which they played the roles of different stakeholders (Fig. 7.1). The workshop theme was solutions to human-bear conflicts in a mountain village of Ibaraki city (the location of Ritsumeikan University Osaka Ibaraki Campus) in Osaka Prefecture. The instructor (I) set out a hypothetical situation whereby a village is faced with human-bear conflicts, specifically, increased bear sightings and agricultural damage caused by bears. While this situation is imaginary, black bears have been spotted recently in mountainous and rural areas of Ibaraki city. As the bear population in northern Osaka increases, local residents face the very real prospect of bear conflicts in the near future. Fictional roles in the workshop included facilitators, farmers (local residents who have experienced agricultural damage by bears), researchers (ecologists, sociologists), animal right association staff, hunters, business people (selling electric fences, traps, etc.), and government officials (city government and Ministry of Environment). For 2 weeks, each student studied both their own role and the overall situation regarding human-bear conflicts in Japan and worldwide. They used this research to explain their situation and demands during the workshop. When everyone had presented, the group was then challenged to reach a consensus about how best to solve human-bear conflicts. Some students were also appointed to a workshop evaluation committee, which assessed both the workshop and the individual roles. The committee presented their findings the following week, including suggestions for improving the effectiveness of the workshop.

7.3 Meaning of the Human Dimensions Course Based on Student Feedback

[For more information about the detailed results of this educational study, see Sakurai 2016.]

Teaching a course, for me, is an ongoing process of revising and polishing in response to student feedback. To increase the course's educational impact, the instructor needs to stay abreast of developments in the field (e.g., actual rural communities in Japan) and ensure that these issues are reflected in the course content. The instructor also needs to pay attention to students' reactions in class, during discussions and in reports, to assess whether the course content is being assimilated and understood.

For my course at Chiba University in 2014, students submitted reports every 1 or 2 weeks stating how they felt about the classes and what they had learned. The contents of these reports are explained in Table 7.2. In the first part of the course (from weeks 2–6), most of students mentioned learning about wildlife issues and environmental problems (e.g., "I didn't know that wildlife issues are such big a problem in this country"). At the end of the course (from weeks 12–14), many referred to the importance of communication (e.g., "I learned that, when conducting interview, it is important to formulate good questions so that the interviewee can understand"). This seems to reflect the contents of the course. The first part was mainly about basic environmental and wildlife issues, while the latter part covered communication

Table 7.2 Contents of reports submitted by students (from the 2014 course at Chiba University) [based on Sakurai 2016]

Weeks in which reports were assigned	Number of reports submitted	Learning about wildlife issues/ environmental problems	Learning through discussions with students from different countries/cultures	Learning about communication
Week 2	10	8	9	0
Week 3	9	8	2	0
Week 4	10	8	1	0
Week6	8	8	2	1
Week7	7	5	0	0
Week 9	11	8	5	0
Week 10	6	5	2	1
Week 12	6	1	3	4
Week 13	8	1	1	4
Week 14	7	4	4	7
Total	82	56	29	17

and survey skills, including interviews and workshops. Several students consistently wrote about the benefits of discussing issues with others from different countries (e.g., "I learned that in other countries, such as the US, hunting is popular among citizens"). The 2014 course at Chiba University included 12 students (7 Japanese, 2 Americans, 1 Australian, 1 Finnish, and 1 Thai). The culturally diverse learning environment gave students a good opportunity to broaden their perspectives.

Comments in the reports submitted by students revealed what they discovered during the course. A male Japanese student commented: "According to H (American female student), people who live in the town use the monitoring system to check whether bears come to the human areas or not (in the US). Residents who have enough time to check this system take turns to be on duty. They can take action as soon as they find out where bears are coming from. However, I think this might be expensive. In Japan, many villagers put fences around their fields to protect their crops and domestic animals. However, wild animals can break these fences easily and these interventions may not be so effective."

A female American student questioned how wildlife control measures – hunting, in her country – could potentially be applied in Japan to mitigate human-wildlife conflicts. "I didn't realize that bears, boars, and deer are such a big issue in Japan. In my state (Wisconsin), we have bears, but there aren't that many issues with bears because a majority of people who live in the country or the woods have weapons (i.e., guns) to protect themselves. It's not only hunters. The people who have guns don't shoot the animals because it is illegal unless the animal is causing harm to them. But the guns are meant to scare the animal, so shots are fired to scare the animal with the noise. Also, in my state, there are many hunters, so that also helps control the populations of various animals such as deer, bears, wolves, turkeys, etc. I think maybe the Japanese government or cities need to in some way entice people to be hunters."

When hunting was discussed in class in terms of wildlife management and tradition, several Japanese students seemed to change their perceptions of the issues after discussing with students from the USA, Finland, and Australia, where the practice is still widespread.

> I thought hunting was a bad activity. I didn't understand the importance of hunting. But now, I understand the role of the hunter. To stop the decrease of hunters in Japan, I think we need to know about the role of hunters first. (A comment by a female Japanese student)

> I learned that human hunters play an important role in Satoyama. (A comment by a male Japanese student)

Meanwhile, some students felt that they enhanced their communication skills by interacting with other students from different cultures and countries. For example, in the interview exercise, some realized the difficulty of communication, as well as how interviewees' thoughts and answers could be very different from their own.

> I was surprised that other people from my group ended up having trouble with the same question. We realized that subtle differences in word usages create language barriers, such

as the one that we experienced. These differences stem from cultural differences and general differences in the way we think due to influences such as culture and society. (A female American student)

During the last class, I felt that writing good interview questions was really difficult. When a foreigner interviews a Japanese person, they should use words that are easy to understand. If you cannot understand the meaning of the question, the interview itself is not meaningful. (A female Japanese student)

In the workshop at the end of the course, students learned the difficulty of reaching consensus among stakeholders. They were also made aware of the significance of holding workshops and the importance of listening to differing opinions, as well as encouraging others to speak up.

When the workshop began, I was thinking, 'Can this really work, because we don't have any experience of being government officials, animal right association staff, farmers, facilitators or even researchers?' After we started in the role of facilitators, however, I began to think that maybe we could do it after all. When the workshop was over, I was amazed at how well we managed. We even achieved a consensus, which was unanimously approved. This workshop activity was really nice. Actually playing the roles gave me and maybe others some insight into new ways of looking at the situation. (A female Finnish student)

I realized that when certain people are discussing issues, others are unable to join in. In situations like this, the facilitators play an important role. They should ask less active members to speak up, improving the discussion. Also, I realized that when you listen to someone telling a story, you should wait until he or she has finished. (A Japanese female student)

Reports by some students explained how they integrated what they had learned on the course into further discussion about how human-wildlife issues related to their own lives.

I got interested in wildlife issues in Japan. Before taking this class, I didn't know much about wildlife in Japan because I live in the city. But would I get a deeper understanding of wildlife if I lived in the rural area? I would not. For example, I sometimes go to Nagano to see my relatives. They live in a mountainous area, where I see wild birds and animals, like kites. My relatives are farmers, so they take actions to prevent wildlife from eating their crops. However, other people do not understand their concerns. Many people who don't understand wildlife conflicts, including me, like to use the phrase; "Save wildlife!" Of course, saving wildlife is very important. We should build a stronger connection between people who are trying to prevent wildlife damage. To save wildlife, we should first know the facts. (A male Japanese student)

7.4 Significance of Human Dimensions Education

Human dimensions of wildlife management is an academic discipline that requires learners to understand not only the background of why and how human-wildlife conflicts occur but also ways to mitigate these conflicts. Managers and researchers contemplating solutions to human-wildlife conflicts need to be aware of multiple concepts and aspects, including related law and policies, theory and approaches to decision-making and governance, and issues pertaining to biodiversity conservation (Decker et al. 2012). The extent to which these themes and concepts can be covered during a 15-class course, however, is somewhat limited.

The course described in this chapter introduces students to the basics of the natural environment, biodiversity, and ecological concepts. However, these topics are only covered in the opening weeks. More significantly, perhaps, students learned to recognize the difficulty of incorporating differing beliefs and opinions and reaching consensus for a group of people with divergent stakes and interests. This was one of the main aims of the course. To this end, I put a lot of time and effort into organizing interview exercises, debates, workshops, etc.

Human-wildlife conflicts, including agricultural damage and occasional human casualties, are a worldwide issue (Manfredo 2009). As I explained in Sect. 2.1, human-wildlife conflicts, in many instances, are actually human-human conflicts, for example, conflicts between local farmers who have suffered agricultural damage and thus would rather eradicate the responsible species and governments that need to design and implement policies centered on conservation and management. Similarly, conflicts exist between animal rights association and hunters. To solve the issues, it is important that stakeholders communicate about the issues and collaborate to mitigate the conflicts.

Through this course, I sought to provide a learning environment in which students from diverse countries and cultures were able to discuss and understand the similarities and differences in global human-wildlife conflicts and consider ways to mitigate the issues (e.g., hunting) in different parts of the world. Additionally, students learned how difficult it is to reach consensus, even in a hypothetical scenario. By developing their communication skills, they learned how to reach agreement via a participatory process – an essential tool for any prospective human dimensions researcher seeking to carry out field work or implement projects on the ground (Jacobson and McDuff 1998; Decker et al. 2012).

The course could be improved in several areas. For example, visiting the sites where human-wildlife conflicts occur in person, conducting field studies to inspect the damage sites, and interviewing local residents, government officials, hunters, and other stakeholders would provide students with a more realistic impression of the real-life situation. In April 2018, I launched a course at Ritsumeikan University that aims to incorporate these criteria. For 1 year, international students continuously visit the mountainous village at Ibaraki city, where they perform agricultural work with local residents (Fig. 7.2). They also hold interviews and stage field trips to damage sites. Their growth as professionals is deeply linked to the strength of their links to local communities and the situation on the ground (Decker et al. 2012). Students require theoretical understanding, skills, and field experience. They should hence aim to take courses that incorporate all of these aspects – i.e., a combination of lectures, active leaning components, and field work.

To teach human dimensions in wildlife more systematically in Japan, a model curriculum and syllabus that is accessible to instructors from different universities and in different countries should be developed. Such undergraduate-/graduate-level human dimensions in wildlife courses currently exist in the USA. Key elements in

Fig. 7.2 Students engaging in agricultural work with local residents in the mountainous village. Field activities are part of the course started at Ritsumeikan University in 2018

human dimensions education are also explained in several books (e.g., Decker et al. 2001). Meanwhile, more efforts should be made to coordinate human dimensions education in a global context and to examine the effectiveness of existing courses in terms of increasing students' knowledge and skills.

References

Decker DJ, Brown TL, Siemer WF. Human dimensions of wildlife management in North America. Maryland: The Wildlife Society; 2001.

Decker DJ, Riley SJ, Siemer WF. Human dimensions of wildlife management. 2nd ed. Maryland: Johns Hopkins University Press; 2012.

Jacobson SK, McDuff MD. Training idiot savants: the lack of human dimensions in conservation biology. Conserv Biol. 1998;12(2):263–7.

Manfredo MJ. Who cares about wildlife?: social science concepts for exploring human-wildlife relationships and conservation issues. New York: Springer; 2009.

Sakurai R. Implementation and evaluation of a course regarding wildlife issues taught to international students at a university: qualitative analysis of reports submitted by students. Jpn J Environ Educ. 2016;26(1):38–51. (in Japanese with English abstract).

Chapter 8
Potential of the Human Dimensions Approach: From Asia to the Wider World

Abstract Wildlife issues are often said to be global/international issues, and wildlife professionals require a global understanding of these challenges. This chapter first explains the potential of the human dimensions approach in solving human-wildlife conflicts, both in Asia and worldwide. Issues facing Japan such as aging and depopulation will soon be issues that many countries around the world will be forced to address. Studying how human dimensions research could mitigate human-wildlife conflicts in Japan and Asia may hence bear insights that may be applicable elsewhere. In that sense, Japan and other Asian countries could become model cases for introducing human dimensions approaches into wider management strategies aimed at solving human-wildlife conflicts. The second part of the chapter introduces case studies in which the human dimensions approach was used for research into participatory coastal management. The concept of "citizen science" is also introduced, including an explanation of how it could potentially be utilized for better community-based wildlife management.

8.1 Japan and Asian Countries as Model Cases for the Human Dimensions Approach

Wildlife issues are in many ways international matters. A simple example of their global nature would be migration, whereby cranes, butterflies, caribous, and many other species fly and run across national boundaries. Similarly, invasive species arriving in their non-native landscapes are some of the biggest current threats to biodiversity. To conserve and manage wildlife, global perspectives are important for establishing regulations, laws, and policies at both international and national levels. In the current era, any person could be connected to environmental destruction, either directly or indirectly. For example, paper items used daily by citizens in developed countries may be made of trees cut down from tropical rainforests in developing countries. In this context, Decker et al. (2012) state that "wildlife management is a global profession" and that "every wildlife professional needs

R. Sakurai, *Human Dimensions of Wildlife Management in Japan*, Ecological Research Monographs, https://doi.org/10.1007/978-981-13-6332-0_8

educational experiences that contribute to their global understanding and appreciation" (p. 244).

A majority of the research presented in these pages was designed and implemented on the basis of the ideas, theories, and methods of linked to human dimensions of wildlife management that I first encountered at graduate school in the USA. I went on to conduct a series of human dimensions studies that showed the ways in which the human dimensions approach could be applied to Japanese culture to examine cognitive function in respect to wildlife issues, analyze behavioral intentions and actual behavior to prevent damage, and clarify the overall social characteristics of human-wildlife issues in Japan.

In this chapter, I would like to delve deeper and discuss the potential of the human dimensions approach to solve human-wildlife conflicts both in Asia and worldwide. Japan currently faces a severe depopulation and aging crisis. The population of Japan, which has long been decreasing, currently stands at 127 million (2013) and is predicted to decline to less than 100 million in 2050 (Cabinet Office Japan 2014). Predictions are that 1 in every 2.5 people will be 65 or older by 2060 (Cabinet Office Japan 2014). The rate/speed of aging in Japan is one of the highest and fastest in the world (Miller and Spoonman 2015, p. 133). In rural areas such as mountainous villages where much of the research presented in this book was carried out, depopulation and aging will be even more severe. Similarly, as the hunter population decreases in Japan, human-wildlife conflicts in these rural areas will worsen. Studying Japanese wildlife issues and applying human dimensions methods and theories to mitigate of human-wildlife issues can produce data that are both instructive in its own right and, importantly, applicable in other countries. Indeed, many developed countries have begun to face similar problems to those encountered by Japan, such as a shrinking and aging society (e.g., Korea, England, and Italy) and a dwindling hunting population (e.g., the USA). On the other hand, many other Asian countries face the opposite problems, i.e., a rapidly growing human population that poses a threat to wildlife (e.g., India, Indonesia, and parts of Africa). Japan and other Asian countries can thus be viewed as model cases for introducing the human dimensions approaches into management strategies for solving human-wildlife conflicts.

8.2 Human Dimensions for Natural Resource Management and the Potential of Citizen Science

Implementing the human dimensions approach in wildlife management research remains both a learning experience and a process of trial and error. Seeking the optimal human dimensions theories and tools to use in different circumstances is an ongoing challenge. Throughout this book, I have explained how the human dimensions of wildlife management can explain how and why human-wildlife conflicts occur and how this novel approach may offer workable resolutions to the issues. In this final chapter, I will hence explain how the human dimensions approach could be expanded to other fields beyond the immediate scope of wildlife management.

The more research I carry out into human dimensions of wildlife management, especially in fields where damage by wildlife is a real and present threat, the more I recognize that wildlife issues are connected to a range of other social issues, not least the depopulation and aging of rural communities that are frequently linked to urbanization and centralized government policies. Another recurrent theme is that wildlife management and natural resource management are closely related. Indeed, wildlife itself can be categorized as a "natural resource," and wildlife management frequently refers to questions of how to manage forests, the *Satoyama* ecosystem, and others. Therefore, although the main topic of this book was the human dimensions of "wildlife management," the ideas, approaches, tools, and theories contained herein could equally be applied to other fields beyond the framework of human-wildlife conflicts.

For example, my wildlife management research has led me into the field of sustainable coastal management in Japan. In Sect. 3.3, I introduced *Satoyama* (*Sato* means villages and *Yama* means uplands in Japanese) – the typical landscape around residential areas including woods, paddy fields, and rivers. *Satoyama* serves as a buffer zone between human and wildlife habitats, as well as being a rich source of biodiversity. Similarly, in many coastal areas of Japan, local fishermen and community residents are engaged in the sustainable management of fish and other marine resources. These areas where both biodiversity and production, including the fish harvest, have been increased through human involvement are called *Satoumi* (*Sato* means village and *Umi* means ocean in Japanese) (Yanagi 2010). However, as for Japan's mountainous villages, many fishing hamlets in coastal areas are facing issues of depopulation and aging. Similarly, the number of fishermen able to manage and monitor fish and other aquatic resources is decreasing. Understanding local residents' perceptions of coastal management, as well as their willingness to get involved both directly and indirectly in coastal conservation issues, is thus key to any sustainable management ambitions.

Drawing on ideas from the human dimensions approach, I used social-psychological theories to analyze local residents' attitudes and behavioral intentions toward coastal conservation. Similarly, I used methods from social science to reveal factors affecting local residents' willingness to conserve the coastal area (Table 8.1) and the conceptual framework by which they gain their sense of place (Fig. 8.1). The results led to several suggestions for future policies and outreach programs to enhance public support and participation. In Sect. 4.5, I described a study that examined how best to evaluate a community education program, and similar research was conducted on this occasion in the field of coastal management. The target here was a marine education program for local junior high school students, and the aim was to examine how this program affected students' attitudes and behaviors toward the conservation of the local marine environment (Fig. 8.2).

Basic ideas and goal of human dimensions studies, i.e., utilizing social science theories and tools to solve real-life problems, could be applied to different fields, as shown in the present coastal management research.

Table 8.1 Example results of the human dimensions studies approach applied to coastal management; multiple regression analysis of residents' willingness to invest their time and effort to conserve the coastal area around Shizugawa district in Miyagi Prefecture

		B	p	VIF	R^2	Adjusted R^2
Biophysical variables	I want to know more about the coastal area wildlife	0.209	<0.001	2.418	0.583	0.580
	I am strongly connected to the land and natural features that make up the coastal area	0.150	<0.001	2.020		
	I hike, go bird-watching, look at wildflowers, and observe nature around the coastal area	0.136	<0.001	2.028		
	I like various outdoor activities that are available in and around the coastal area	0.119	<0.001	2.194		
	I think the coastal area's biodiversity is important	0.112	<0.001	2.135		
	I think the coastal area's wildlife is fascinating	0.058	<0.05	2.519		
Psychological variables	I feel like the coastal area is a part of me	0.068	<0.05	2.727		
	I would rather live in the coastal area than anywhere else	0.056	<0.05	2.528		
	I get more satisfaction out of living in the coastal area than living anywhere else	0.054	0.07	2.946		
Attitudes toward the future generation	I want to keep the coastal area in a good condition for the future generation	0.064	<0.01	1.813		
Socio-demographic variable	Living distance from the coastal area (Shizugawa Bay)	−0.040	<0.05	1.053		

n = 1,418 (based on Sakurai et al. 2016)

Currently, "citizen science" is defined as "public participation in organized research efforts" (Dickinson and Bonney 2012) or "engaging the public in a scientific project" (Kobori et al. 2016). It has become increasingly important for understanding environmental changes both locally and at a global level. Citizen science has a long history in the fields of ecology and conservation, and current projects such as Feeder Watch and Christmas Bird Count involving a large number of participants drawn from the general public have revealed important information for bird species conservation (Dickinson and Bonney 2012). In both of these projects, citizens observed the birds around their houses or residential area and reported their findings to the organization in charge of the project via a website.

The majority of discussions about citizen science in the field of ecological science address ways in which people can get engaged in compiling data for species conservation. However, I believe the discipline can be usefully applied to the mitigation of human-wildlife conflicts all over the world.

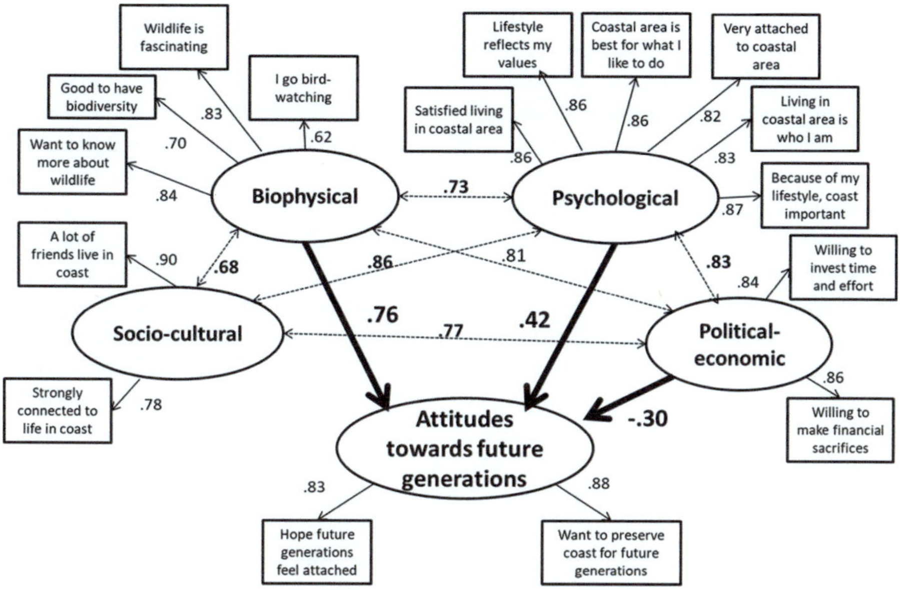

Fig. 8.1 Example results of human dimensions studies applied to coastal management; structural equation modeling based on a survey conducted around Hinase district, Okayama Prefecture. n = 2,851. (based on Sakurai et al. 2017)

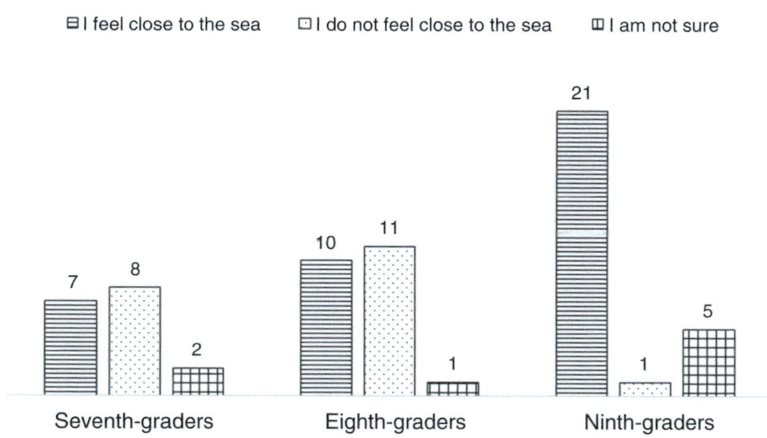

Fig. 8.2 Example results of human dimensions studies applied to coastal management; students' perceptions of a marine education program at a junior high school in Hinase district [this question was specifically asking whether students felt an affinity for Hinase's sea]. (based on Sakurai et al. 2018)

In the course of my human-wildlife conflict studies, many local residents have kindly showed me the damage sites. They knew a lot about the local situation, such as where wildlife appeared, in many cases more than the researchers. In one such case, in a village of Tochigi Prefecture, one resident started to put camera traps around the village to understand and monitor where and how the species appear. He used these data to work out where to place his fences and traps and to establish what types of countermeasures would be effective against the wildlife. He also used the video recordings to warn and educate local residents about the wildlife populations surrounding the villages and to engage them in damage prevention behavior. This person was not a "professional" scientist, and he did not have a PhD in wildlife management. However, his actions contributed significantly to strengthening the local communities' resilience to wildlife damage. There are always limitations present in the numbers of researchers, local government staff, and insufficient wildlife damage budgets, and a local community's capacity to successfully and independently overcome wildlife issues may ultimately depend on these "citizen scientists." In this sense, it could be an effective method for implementing collaborative wildlife management among local residents, researchers, and governments.

Although it is important for human dimensions researchers and practitioners to understand the ecology of wildlife and issues specific to wildlife management, the human dimensions field could be enhanced and developed by association with other related fields and themes. Applying human dimensions tools and ideas to different fields of natural resource management and utilizing disciplines such a citizen science could all be fruitful ways to expand the potential of the human dimensions field.

References

Cabinet Office Japan. Annual report on the aging society, Tokyo, Japan. 2014. (in Japanese). Available from http://www8.cao.go.jp/kourei/english/annualreport/2014/2014pdf_e.html. Accessed July 2015.

Decker DJ, Riley SJ, Siemer WF. Human dimensions of wildlife management. 2nd ed. Maryland: Johns Hopkins University Press; 2012.

Dickinson JL, Bonney R. Citizen Science: public participation in environmental research. New York: Comstock Publishing Associates; 2012.

Kobori H, Dickinson JL, Washitani I, Sakurai R, Amano T, Komatsu N, Kitamura W, Takagawa S, Koyama K, Kgawara T, Miller-Rushing AJ. Citizen science: a new approach to advance ecology, education and conservation. Ecol Res. 2016;31(1):1–19.

Miller GT, Spoonman SE. Living in the environment. Connecticut: Cengage Learning; 2015.

Sakurai R, Ota T, Uehara T, Nakagami K. Factors affecting residents & behavioral intentions for coastal conservation: case study at Shizugawa Bay, Miyagi, Japan. Mar Policy. 2016;67:1–9.

Sakurai R, Ota T, Uehara T. Sense of place and attitudes towards future generations for conservation of coastal areas in the Saotumi of Japan. Biol Conserv. 2017;209:332–40.

Sakurai R, Uehara T, Yoshioka T. Students' perceptions of a marine education program at a junior high school in Japan with a specific focus on Satoumi. Environ Educ Res. 2018; https://doi.org/10.1080/13504622.2018.1436698.

Yanagi T. Creation of Satoumi. Tokyo: Kouseisha Kouseikaku; 2010. (in Japanese).

Index

R. Sakurai, *Human Dimensions of Wildlife Management in Japan*, Ecological Research Monographs, https://doi.org/10.1007/978-981-13-6332-0

MIX
Papier aus verantwortungsvollen Quellen
Paper from responsible sources
FSC® C105338

If you have any concerns about our products,
you can contact us on
ProductSafety@springernature.com

In case Publisher is established outside the EU,
the EU authorized representative is:
Springer Nature Customer Service Center GmbH
Europaplatz 3, 69115 Heidelberg, Germany

Printed by Libri Plureos GmbH
in Hamburg, Germany